Climate Change in America

Ecosystems and Policy

Oliver Brooks

Global East-West. London

Copyright © 2025 by Oliver Brooks.

Global East-West. London.

All rights reserved.

No portion of this book may be reproduced in any form without written permission from the publisher or author, except as permitted by copyright law.

Contents

1. Introduction 1
   Understanding the Climate Crisis in America
2. The Science of Climate Change 21
   A Primer
3. Precarious Balance 39
   The Imperiled Biodiversity of America
4. Vulnerable Populations 47
   Facing Disproportionate Risks
5. Regional Impacts 67
   Case Studies Across the Nation
6. Federal Environmental Policy 85
   A Historical Overview
7. State and Local Leadership in Climate Action 105
8. US Engagement in Global Climate Action 125
   A Comparative Analysis of Republican and Democratic Administrations
9. Balancing Mitigation and Adaptation Strategies 131

10. Future Directions 155
    Navigating Political and Social Complexities

References For Further Reading 177

1
Introduction
Understanding the Climate Crisis in America

The Urgent Need for a Global Lens

The rise of climate change as a defining challenge of the twenty-first century compels scholars and policymakers to adopt a genuinely global lens when crafting responses. The intricate and reciprocal links among environmental, economic, and social systems worldwide mean that no single nation, including the United States, can shield itself from the effects of rising atmospheric greenhouse gas concentrations. Analysing how global processes shape, and are shaped by, U.S. climate strategies is now a scholarly as well as strategic necessity. The architecture of multilateral accords, the design and diffusion of emissions trading markets, and the movement of goods and energy across frontiers all shape domestic decisions, while American actions reverberate across the same systems. This reciprocal dynamic obliges analysts to move past a solely national frame and to study transnational feedback loops—political, technological, and ecological. The history of climate change is, in part, a history of differential responsibilities, with industrialised nations having poured far greater quantities of carbon dioxide into the atmosphere per capita in the past. Contemporary emissions profiles differ still more, as developing economies adopt carbon-intensive pathways alongside cleaner technologies. Thus, principles of equity and fairness resonate across the negotiations, insisting that any U.S. strategy must either lead or transparently account for global responsibilities. Against this backdrop, unilateral signals—however well-intentioned—will remain dwarfed by the emissions of other nations unless they are coupled with diplomatic, financial, and technological cooperation. The reciprocal learning that occurs when nations share policy architectures and technological trials will, in the long run, strengthen domestic resolve and broaden the arsenal of options available.

Active participation in the global climate community, a network of nations, organizations, and individuals dedicated to addressing climate change, enables the United States to adopt proven insights and strategies that strengthen domestic climate resilience and emission-reduction efforts. Simultaneously, the shifting landscape of global climate diplomacy informs the circumstances under which U.S. policies are conceived and executed. Exposure to a broad spectrum of international viewpoints and practices deepens the national conversation on climate change, promoting a more integrated and participatory policy-making process. Framing the climate crisis in a global context reveals the inherent interdependence of all countries in confronting this existential risk. Acknowledging the mutual stakes and obligations that accompany climate action at the planetary level is essential for devising durable solutions and nurturing international solidarity in defence of our shared future.

Historic Emissions: The Weight of the American Footprint

America's carbon legacy is woven into the very fabric of

its economic evolution. From the first European settlements onward, the systematic extraction of resources, coupled with the vigorous spread of commerce and manufacturing, laid the groundwork for extensive greenhouse gas release. The shift from farming to mass production in the late nineteenth and early twentieth centuries triggered rising output from factories, railroads, and power stations. An increasing dependence on coal, and later on oil, intensified the trend. The economic boom following the Second World War accelerated the scale and reach of the American carbon stream; mass production techniques and consumer culture drove elevated energy use and waste generation. As the United States rose to the status of the world's foremost economic power, its carbon output continued to rise, undergirded by urban expansion, demographic growth, and rapid technological change.

The trajectory of historical emissions, therefore, encapsulates intricate socioeconomic processes. Although industrial activities clustered in defined locales, their reverberations extended nationwide, impacting communities, ecosystems, and public health. Populations residing in proximity to industrial perimeters or fossil fuel extraction zones frequently endure disproportionate exposure to pollutants and environmental decline, thereby entrenching social inequities and health disparities. The continuity that binds past emissions to present environmental justice imperatives starkly illustrates the persistent repercussions of the United States' cumulative carbon output.

A robust understanding of how emissions evolved must also be anchored in the decisions of policymakers and the choices of society. Legislative frameworks, coupled with systematic redistributions of energy networks, chart a historical course that today informs both the obstacles and the avenues available for effective climate mitigation. Analysis of temporal emission trends yields not only a recognition of their enduring effects, but also the analytical baseline needed for the design of responsive climate governance. In a period of intensified global consciousness about climate change, this historical lens on American emissions functions as both a reminder of past accountability and a summons to vigilant and ethically responsible environmental stewardship.

Economic and Social Dimensions of Climate Change

Climate change manifests across both economic and social spheres, compelling developed and developing nations alike to reconcile multiple and sometimes conflicting priorities. On the economic front, shifting climate regimes disrupt agricultural output, elevate the frequency and severity of damage to physical infrastructure, and escalate public health expenditures tied to heat-related illnesses and worsening air quality. While the aggregate financial burden is considerable, the distribution of costs is uneven, with vulnerable populations bearing the greatest exposure to both immediate shocks and chronic stressors. These intersecting pressures deepen inequality, undermine formal and informal safety nets, and jeopardise subsistence and market-oriented livelihoods. At the same time, climate-induced variability in crop yields and market access jeopardises global food security, creating the potential for localised scarcities and global price volatility that reverberates through all social strata. The required shift toward low-carbon production and consumption introduces both risks and opportunities that necessitate sector-specific innovation, technological evolution, and managerial reconfiguration. Socially, climate change erodes the stability of communities, disrupts customary practices, and intensifies competition for scarce resources, thereby fuelling conflict in some contexts and migration in others. Marginalised and historically disadvantaged groups experience disproportionate exposure and reduced capacity for adaptation, necessitating targeted resilience-building that accounts for power asymmetries. Awareness of these interwoven economic and social dimensions is indispensable for crafting integrated, durable policy architectures and for engaging a broad coalition of public, private, and civil society actors.

Sustained attention to the interplay between economic, social, and ecological imperatives of climate change requires a commitment to long-range planning, the integration of sustainability into all dimensions of development, and strengthened multilateral collaboration to forge a fair and flourishing global future.

Environmental Degradation: A Comprehensive Examination

Environmental degradation, the deterioration of the environment through depletion of resources such as air, water, and soil, as well as the destruction of ecosystems and the extinction of wildlife, consists of a complex web of processes that collectively undermine the integrity of natural systems and the resources they provide. Deforestation, habitat fragmentation, the ubiquity of pollutants, and the progressive warming of the atmosphere exemplify the ways human enterprise magnifies ecological risk. This section delves into the multiple, interlinked aspects of degradation to illuminate the consequences that confront both current and succeeding cohorts.

Among the threats most urgently demanding attention is the accelerated collapse of biodiversity. Population growth, land-use change, and the commodification of ecosystem services are eroding the conditions that sustain species variability. Once-stable ecological equilibria are undergoing disruptive reorganisation, generating ripple effects that destabilise food webs, alter hydrological regimes, and compromise the resiliency of entire biomes. The erosion of genetic and species-level variation also diminishes humanity's potential to discover novel medical therapies and agricultural innovations, as irreplaceable genetic reservoirs are irretrievably lost.

A further essential dimension of environmental degradation is the contamination of air, water, and soil. Industrial production, agricultural expansion, and accelerated urban development have led to the discharge of hazardous substances into ecosystems, aggravating air and water quality crises while accelerating soil degradation. These contaminants jeopardise both public health and wildlife, and they impose enduring risks on the availability of critical resources such as potable water and arable soil.

The transformation and outright destruction of ecosystems through large-scale deforestation and unconstrained urban sprawl also produce far-reaching effects on climatic and meteorological systems. Forests function as substantial carbon sinks, sequestering and stabilising atmospheric carbon dioxide. When they are systematically removed, the balance of greenhouse gases in the atmosphere is disturbed, accelerating the process of global warming and the destabilisation of environmental systems.

Effective responses to environmental degradation must therefore be multidimensional, integrating sustainable resource stewardship, habitat preservation, and coordinated policy action at local, national, and international levels. By dissecting the intricate interdependencies that underlie environmental decline, societies can develop and apply robust strategies that reduce immediate harms while safeguarding the ecological fabric of the planet for the benefit of subsequent generations.

Policy Responses: A U.K. Retrospective

The trajectory of U.K. environmental governance demonstrates both institutional resilience and political complexity. Beginning in 1970 with the creation of the Environmental

Protection Agency, the country institutionalised a regulatory framework anchored by the Clean Air and Clean Water Acts, which established mandatory control of emissions, the preservation of aquatic ecosystems, and the responsible management of renewable and nonrenewable resources. Over subsequent decades, the U.K. also mobilised in multilateral venues, including the ratification of the Paris Agreement, pledging to contribute to a coordinated global strategy to mitigate climate change. Yet while these milestones signify an overarching governance commitment, their operational record reveals uneven outcomes. The intricate division of regulatory authority among federal, state, and local jurisdictions, compounded by active lobbying from industrial sectors, environmental NGOs, and the scientific community, has produced a policy environment marked by both incremental gains and pronounced regulatory backsliding. By tracing statutory enactments, judicial interpretations, and administrative shifts, the present analysis seeks to elucidate the converging dynamics that produced the contemporary climate emergency, spotlighting decisive legislative moments, the mobilisation of entrenched and emergent constituencies, and the enduring challenge of aligning short-term political incentives with long-term ecological imperatives.

A critical study of earlier climate policies—framed by their political, social, and economic contexts, their varied achievements, and their weaknesses—yields instructive perspectives on contemporary options for meaningful action. The U.K. record, reviewed over decades, underscores that effective, durable climate governance depends on an iterative, cooperative, and evidence-driven approach. The balance of continuity and responsive adaptation must guide present and future efforts. Subsequent chapters will analyse how legal and institutional architectures have influenced observable environmental change, distilling durable lessons from historical patterns to strengthen current strategic responses to the climate crisis.

Scientific Consensus: Uniting Behind the Data

The scientific consensus concerning climate change attests to the systemic and collaborative character of global scientific inquiry. Evidence collected across climatology, oceanography, and atmospheric sciences converges to identify human activity as the dominant force shaping the current trajectory of the Earth's climate. Through sustained observation, rigorous data gathering, and refined modelling techniques, researchers have explicated the complex feedbacks that govern planetary climate dynamics.

Critical data, including the accelerated rise of atmospheric greenhouse gases, ocean thermal expansion, and the increasing frequency of extreme meteorological phenomena, compel an immediate and sustained response. The Intergovernmental Panel on Climate Change (IPCC), an assemblage of several thousand scientists and country delegates, continues to produce comprehensive assessment reports that translate the scientific data into actionable knowledge, delineating both the magnitude of human influence and the strategic options for mitigation and adaptive resilience.

The convergence of findings from diverse lines of investigation fortifies climate science and counters baseless dissent. Stringent peer review and open methodologies validate and corroborate scientific claims, thereby informing policy deliberation at domestic and global scales. The consensus among climate scientists concerning the reality and seriousness of anthropogenic climate change offers a stable platform for developing regulatory frameworks and for raising public consciousness.

In parallel, improvements in observational sensors, remote sensing, and computational power have enhanced climate modelling, yielding projections of future states that are increasingly precise. Such predictive skill is indispensable for quantifying exposure to hazards and for crafting resilient strategies that protect critical infrastructure, natural systems, and at-risk populations. Ongoing collaboration among scientists, research consortia, and public authorities persistently sharpens the empirical underpinnings of climate action, thereby justifying and expediting effective policy responses.

Driven by a resolute dedication to data-centred scholarship and cross-disciplinary coordination, the global research community asserts that immediate, coordinated efforts are essential to confronting the complex crises generated by changing climates. By leveraging the cumulative insight generated over recent decades of rigorous investigation, societies can navigate a resilient, equitable future, ensuring the planet's ecosystems endure for the benefit of generations to come.

The Collective Challenge of Public Understanding

Public understanding is a decisive factor shaping climate dialogue. Although researchers concur on the reality and consequences of climate change, citizens manifest diverse beliefs and misunderstandings. This segment critically examines the determinants that condition public views, clarifies widely circulated fallacies, and advances methods that can cultivate a more nuanced and scientifically grounded civic conversation.

Central to public understanding are the underlying ideological and epistemic frameworks that individuals bring to the question. Preconceived attitudes toward science, governance, and economic systems powerfully condition assessments of climate dynamics. Political identities, in particular, often determine the degree of alarm or disengagement expressed, with partisan divisions either intensifying or diminishing engagement with climate science and policy.

Moreover, the way in which the media presents and frames climate-related subjects decisively shapes public comprehension. The choice of accurate, peer-reviewed findings over misinformation or contrarian viewpoints, as selectively broadcast by some outlets, can tilt public attitudes in significant ways. A systematic appraisal of the media's contribution to public misconceptions must therefore be undertaken, alongside the development of strategies to elevate balanced, evidence-centred journalism.

Erroneous beliefs about climate change are pervasive and merit rigorous analytical untangling. Illustrative errors include the claim that natural temperature variances undermine evidence of human influence, or the false equivalence that climate experts are privately uncertain. Countering such beliefs demands the continual re-articulation of the extensive evidence linking fossil fuel combustion, deforestation, and other anthropogenic activities to the warming trend, as well as the consistent endorsement of the crisis's reality and gravity by virtually the entire scientific community.

Creating a public capable of informed and constructive engagement with climate policy necessitates a sustained commitment to scientific literacy and critical reasoning. Instruction about the climate system's intricate dynamics, the protocols that underpin climate measurements, and the socio-economic ramifications of policy delay equips citizens with the knowledge to recognise the crisis's immediacy and to advocate for robust solutions.

Furthermore, it remains essential to convene a broad array of stakeholders—public agencies, private enterprises, research institutions, and civil society organisations—in open and productive dialogue. When we create settings in which diverse views are actively heard and respected, we foster a shared comprehension of the climate emergency and lay the foundation for a unified, deliberate response.

Ultimately, clarifying public attitudes and correcting widespread misperceptions of climate change are indispensable for cultivating widespread support for effective climate policies. By carefully unpacking the complex dimensions of the problem and striving to mend societal rifts through engaged and evidence-based conversation, we may ignite the cross-sectoral momentum required to reduce greenhouse gas emissions and to strengthen resilience in the face of global environmental change.

The media plays a pivotal role in shaping how the public perceives and understands the climate crisis. As a primary information conduit for vast audiences, media organisations can steer public sentiment, individual behaviour, and the formulation of climate-related policies. Yet, wielding such influence involves navigating a landscape marked by complexity and competing pressures.

Central to the media's function is the obligation to deliver precise and proportionate accounts of climate science and its wider consequences. Professional journalism rests on rigorous fact-checking, dependency on authoritative sources, and the conscious avoidance of sensational framing or partisan divisions. By maintaining such scholarly rigour, the media can cultivate a more informed citizenry and nurture deliberative exchanges on the measures required for effective climate action.

Equally significant is the way in which journalists frame climate-related narratives. Choices regarding terminology, visual content, and story progression can decisively shape how audiences comprehend and emotionally react to the crisis. When climate change is rendered as an obscure or future concern, its immediacy to daily life is easily diminished; conversely, anchoring the crisis within personal, humanised vignettes and spotlighting plausible, inclusive solutions can elevate public investment, affective engagement, and, ultimately, the political will for change.

Journalists and media organisations must consciously accept the dual responsibility of educators and advocates for informed environmental citizenship. By facilitating conversations rooted in scientific consensus and revealing the intricate ties among environmental, social, and economic dimensions of climate change, the media can close knowledge gaps and galvanise collective action. Employing an expanding suite of communication platforms—documentaries, podcasts, and interactive digital resources—permits outreach to varied audiences, cultivating a climate discourse that is both sophisticated and inclusive.

Simultaneously, the media confronts formidable headwinds in its coverage of climate change. Complex political and economic interests, persistent misinformation, and a distrustful public complicate the framing of stories. Organisations must also resist the temptation to prioritise sensational engagement at the risk of distorting scientific accuracy. The prevalence of echo chambers and the deep polarisation of the climate debate further complicate coverage, requiring a deliberate commitment to conversations that bridge divides and prioritise evidence-based deliberation over allegiance to tribal narratives.

In summation, press institutions are central to structuring public discourse on climate change and galvanising societal responses to the crisis. By adhering to the tenets of responsible journalism, by employing strategic framing techniques, and by fostering robust, evidence-based debate, the media can motivate individuals, communities, and legislators to confront climate change's manifold perils and its potential for transformative progress.

Key Stakeholders: Government, Industry, and Citizens

In the multifaceted arena of climate policy, the contribution of core stakeholders—government, industry, and the general citizenry—remains indispensable for translating ambition into tangible results. This portion of the text examines the distinctive yet interdependent functions each constituency bears as the United States confronts the climate emergency.

Government entities, spanning federal, state, and municipal levels, are charged with crafting and executing the regulatory architecture that undergirds climate mitigation and adaptation. At the national tier, agencies such as the Environmental Protection Agency and the Department of Energy establish binding and incentive-based frameworks that govern pollution control and the transition to low-carbon energy systems. Additional authority resides with state, regional, and local authorities that, increasingly, are pioneering market mechanisms, peer-reviewed resilience strategies, and cross-sectoral collaborative programmes to accelerate emission reductions and to enhance societal resilience to extreme weather events and sea-level rise.

At the same time, multiple industries continue to alter the climate, both through upfront greenhouse gas releases and the leverage they possess to pioneer greener practices. Corporate actors are uniquely positioned to implement sweeping environmental corrections, whether by funding clean energy, curbing overall emissions, or committing to science-based reduction targets. Moreover, embedding sustainability through every node of the supply chain and prioritising corporate accountability can foster a durable, climate-safe society.

Yet, the decisive agent in combating climate change remains the informed individual. Recognising the environmental footprint of everyday habits allows citizens to choose products, services, and lifestyles that reinforce planetary boundaries. Responsible citizenship manifests in energy-efficient choices, loyalty to low-carbon suppliers, and active support for climate-oriented legislation. Neighbourhood clean-ups, local climate-fair ordinances, and shared-resources cooperatives further demonstrate that civic-level organising can trigger policy and market transformations.

Acknowledging the unique advantages of public policy, corporate creativity, and individual agency reveals that durable decarbonisation hinges on their seamless interaction. This section has shown that each sphere supplies a crucial ingredient, and only their synchronised and mutually reinforcing mobilisation can secure an equitable and climate-resilient America.

Roadmap to Addressing the Climate Crisis

Confronting the climate crisis in the United States demands an integrated roadmap capable of mobilising government, industry, and civil society alike. Given the crisis's complex, interlinked causes and consequences, an effective response must braid together mitigation, adaptation, and development efforts. The following structured guide delineates key trajectories for advancing a resilient and equitable low-carbon future.

1. Policy Framework: Meaningful climate action depends on coherent policy architectures that articulate enforceable emission reduction milestones, catalyse renewable energy integration, and reward sustainable practices in every sector. Coordinated action across federal, state, and municipal levels is essential to create consistent, scientifically informed measures that can shield populations from both the immediate and the chronic hazards posed by a warming climate.

2. Innovation and Technology: Accelerating the pace of technological advancement is a non-negotiable pillar of crisis response. Continued public and private investment in clean energy systems, carbon management technologies, and regenerative agricultural practices can curtail greenhouse gas releases while stimulating high-quality employment and resilient supply chains. Sustained support for re-

search, development, demonstration, and deployment will be critical.

3. Education and Awareness: Equipping the public with the necessary knowledge is a catalyst for durable change. Structured educational campaigns that illuminate climate science, sustainable consumption, and ecosystem stewardship must be designed to reach diverse audiences and empower households, businesses, and communities to adopt transformative practices. Informed constituencies are also more likely to support and enforce ambitious, equitable climate policies at all levels of governance.

4. Corporate Responsibility: Engaging corporate actors to adopt sustainable operating practices, shrink their carbon footprints, and channel investments into green technologies remains a foremost priority. Corporate social responsibility must now be calibrated to climate targets, driving a systematic transition toward low-carbon operating models and sustainable supply chains.

5. Community Engagement: Enabling local populations to develop and execute grassroots resilience and adaptation strategies is essential. Community-directed efforts, including reforestation, creation of urban green corridors, and investment in climate-resilient infrastructure, strengthen adaptive capacity and simultaneously reinforce social cohesion.

6. International Collaboration: Acknowledging the global nature of climate risk, the United States must deepen multilateral engagement. Diplomatic outreach, dissemination of

scientific and technical expertise, and mobilisation of financial resources in concert with other governments are critical to managing issues that transcend national borders.

7. Sustainable Development Goals: Cross-fertilising climate action with the full set of sustainable development goals amplifies mutual gains. By treating social equity, economic viability, and environmental stewardship as mutually reinforcing imperatives, policymakers can craft pathways that are inclusive and sustainable in equal measure.

Traversing these integrated strategic avenues, the United States can methodically confront the climate challenge while architecting a resilient, low-carbon tomorrow. Coordinated endeavour among public, private, and civic institutions will be indispensable to converting this ambitious but attainable vision into an enduring reality.

2
The Science of Climate Change
A Primer

Foundations of Climate Science

Climate science is an interdisciplinary enterprise that integrates principles from atmospheric science, oceanography, geology, biology, and related fields. Its primary aim is to elucidate the complex interactions among the atmosphere, oceans, land surfaces, and biosphere in order to identify the governing processes that define the Earth's climate system. Mastery of these foundational mechanisms is indispensable for evaluating the multidimensional character of contemporary climate change and its projected consequences.

The Greenhouse Effect: Mechanisms and Implications

The greenhouse effect constitutes a central regulatory mechanism in the Earth's climate system, determining the planet's energy balance. The process is initiated when incident solar radiation strikes the Earth's surface, where it is absorbed and subsequently re-emitted as infrared radiation. A fraction of this infrared energy is absorbed and re-emitted by atmospheric trace gases, including carbon dioxide, methane, and water vapour, thereby forming a thermal "blanket" that reduces the rate of energy loss to space. This net retention of heat stabilises surface temperatures and is a prerequisite for the biospheric processes that sustain life.

Anthropogenic activities, chiefly the combustion of fossil fuels and the large-scale clearing of forests, have markedly augmented the natural greenhouse effect by introducing a greater volume of greenhouse gases into the atmosphere. This enhancement of the greenhouse effect is the principal driver of global warming and the associated changes in the Earth's climate system. The consequences of this heightened atmospheric greenhouse loading are extensive, influencing multiple interconnected Earth subsystems.

The most immediately observable consequence is the systematic rise in global mean temperatures. This thermal upward trend is implicated in the accelerated retreat of polar ice sheets, the comparatively rapid elevation of global sea levels, and an increasing frequency and severity of extreme meteorological events. Accompanying these temperature trends are reconfigured precipitation regimes and notable translocations of climatic zones, phenomena that jointly disrupt natural ecosystems, agricultural productivity, and the availability of freshwater supplies.

In addition, the perturbation of the balance between incoming solar insolation and the outgoing longwave radiation disturbs the planetary energy budget. This imbalance has repercussions for the thermohaline circulation in the oceans and alters atmospheric circulation cells. The resultant changes in the latitudinal distribution of heat and momentum produce downstream weather alterations that compound the hazards posed by an already changing climate.

Understanding the greenhouse effect and its implications is not just important; it's urgent. This knowledge is vital for legislators, researchers, and the public. It's the key to drafting sound policies for emission cuts, planning for the consequences that are already locked in, and negotiating effective global accords. A clear view of the multiple and compounding effects of a stronger greenhouse effect also highlights why rapid action is essential and why moving toward a sustainable, low-carbon economy cannot be delayed.

In brief, a thorough study of the greenhouse effect reveals the interdependence of Earth's physical and biological systems and the critical need for cohesive human action to limit human-driven climate change. It's not just a scientific concept; it's a call to action for all of us to take responsibility for our actions and work towards a sustainable future.

Natural vs. Anthropogenic Factors

A nuanced appraisal of how natural and human-induced factors interact is indispensable for a full understanding of contemporary climate dynamics. Natural drivers include predictable shifts in solar output, volcanic eruptions, and intrinsic variations in oceanic and atmospheric circulation patterns. While these drivers have generated climate variability over geological epochs, producing lengthy arcs of warming or cooling, they cannot account for the rapid temperature rise observed since the nineteenth century. The human factor, or anthropogenic driver, is characterised by the extraordinary pace at which industries, agribusiness, and land-use changes have mobilised carbon dioxide, methane, and nitrous oxide into the atmosphere. This surge of greenhouse gases has strengthened the natural greenhouse effect to such an extent that it now dominates the energy imbalance of the Earth. Quantitative assessment of these competing influences requires sophisticated global climate models coupled with wide-ranging observational data. Results from these efforts converge in demonstrating that the current trajectory of warming is largely attributable to human agency. The task of parsing the simultaneous and often nonlinear processes by which natural variability and anthropogenic emissions interact remains a vital and open line of inquiry in climate science.

Identifying these differences is crucial for designing effective mitigation and adaptation measures against the impacts of climate change. When this broader awareness is embedded in policy formulation and public conversation, communities can work more intelligently toward a sustainable and resilient relationship with the evolving climate system.

Historical Climate Patterns and Anomalies

Scientific advances in climate research continue to refine our understanding of the Earth's long-term climatic evolution and the anomalies that punctuate it. Multi-proxy paleoclimatic investigations built upon ice-core chemistry, dendrochronology, varves, and extrinsic documentary data yield a stratified record of past environmental states. Fluctuations in mean temperatures, hydrological cycles, and rare-event frequencies—documented independently of the industrial age—signal both gradual change and punctuated extremes. Millennial-scale indicators disclose episodes of brief warming and of sustained thermal equilibrium, thereby underscoring the intrinsic multistability of the climate system. 'Quasi-stable states' refer to periods of relative stability in the climate system, where the system is not in a state of rapid change. Through systematic interrogation of standardised and site-specific data, researchers have identified recurrent oscillations, notably the Medieval Warm Period and the subsequent Little Ice Age, that underscore Earth's ability to transition among these quasi-stable states well before the onset of a detectable anthropogenic signature. Comparative integration of these reconstructions supports the influence of both external forcings—orbital, volcanic, and solar variations—and internal redistributions of heat and moisture, delineating the complex, multi-scale mechanisms that have forged the climatic patterns we now seek to project forward in time.

Understanding historical climate patterns is not just about the past; it's about preparing for the future. Recent reconstructions have illustrated how volcanic eruptions, variations in solar output, changes in oceanic circulation, and other natural processes have influenced historical climate anomalies. This knowledge provides important context against which present climatic fluctuations can be assessed. Investigation of these records has further revealed episodes of abrupt climate shifts, highlighting the nonlinear character of the climate system and alerting researchers to the pos-

CLIMATE CHANGE IN AMERICA

sible range of future responses. A rigorous understanding of these historical anomalies thus constitutes an essential baseline that informs climate models and projections, which in turn guide present-day mitigation and adaptation policies. Continuing to probe the past climate record enhances our preparedness for the accelerating pace of current climatic changes.

The examination of climate change reveals that carbon dioxide and the suite of greenhouse gases provide the essential framework for analysing how human activity influences the planet. Methane, nitrous oxide, and various fluorinated substances complement carbon dioxide in their capacity to absorb and re-emit infrared radiation, thereby enhancing the greenhouse effect. While each gas contributes differently, carbon dioxide is of primary concern because of its volume and its long-term stability in the atmosphere. Since the dawn of the industrial era, the widespread burning of coal, oil, and natural gas has markedly elevated carbon dioxide concentrations, initiating a sequence of climatic alterations. Concurrently, widespread deforestation and shifting land uses have destabilised natural reservoirs of carbon, converting once-reliable sinks into additional sources of the gas. The study of these gases thus transcends theoretical interest; it reveals the fundamental mechanisms that govern current warming and guides the transition to a decarbonised future. An effective response to this complex challenge must therefore integrate regulatory frameworks, the pursuit of cleaner technologies, and the mobilisation of communities at local and global scales.

The collective commitment to lowering greenhouse emissions, accelerating the transition to renewable energy, and advancing carbon capture strategies represents our clearest path to curbing the climate crisis. Responding to the rise of greenhouse gases with purposeful and timely action is non-negotiable, as their influences transcend borders and endure for centuries. A complete assessment of carbon dioxide and allied gases empowers us to respond with precision and to secure a climate that is equitable, resilient, and sustainable for present and future generations.

Climate Models and Forecasting Techniques

Climate models and advanced forecasting techniques serve as indispensable lenses through which we explore and predict the intricate workings of the Earth's climate system. These numerical frameworks couple atmospheric chemistry, oceanic circulation, cryospheric processes, and terrestrial systems to reproduce the climate's temporal evolution. Grounded in the conservation of mass, energy, and momentum, they translate empirical observations and theoretical principles into projections of how anthropogenic influences are likely to manifest. By simulating the interactions of myriad variables, the models articulate a spectrum of potential futures in direct response to chosen emissions pathways and the suite of adaptive and mitigative policies enacted.

A critical component of climate modelling is the validation stage, during which simulated results are juxtaposed with observational datasets to gauge their fidelity and robustness. This progressive comparison illuminates areas needing refinement, thereby augmenting predictive power over successive iterations. Concurrently, ensemble techniques, which generate a suite of simulations differentiated by minor perturbations of initial states, permit a quantification of uncertainties that accompany decadal and century-scale climate projections.

Recent enhancements in computational capability permit greater spatial resolution and greater sophistication in representation, empowering scientists to resolve smaller-scale processes and their interlinkages with the climate system. Coupled atmosphere-ocean configurations, for example, systematically incorporate the reciprocal responses of the atmosphere and the sea surface, yielding more credible characterisations of irregular yet recurrent modes, including the El Niño/Southern Oscillation.

Coalescing natural and social science, renewed emphasis on embedding socioeconomic drivers within climate models acknowledges the co-evolution of human behaviour and environmental perturbation. Within this frame, integrated assessment frameworks synthesise physical climate projections with economic trajectories, demographic evolution, and regulatory scenarios, thereby probing the efficacy and cost of diverse mitigation and adaptation pathways. Such cross-disciplinary modelling is indispensable for grounded policy formulation and for calibrating systematic responses to the challenges posed by climate change.

Despite the significant progress made by climate models in elucidating the dynamics of the Earth's climate system, they remain encumbered by fundamental uncertainties and limitations. Granular phenomena—such as cloud microphysics, aerosol-cloud interactions, and complex feedback loops—introduce nonlinear uncertainties that challenge the reliability of multi-decadal projections. Continual refinement of modelling frameworks, supplemented by interdisciplinary dialogue among international climate science communities, is essential for systematically narrowing these uncertainties and enhancing the fidelity of future climate trajectory assessments.

Empirical Observations: Contemporary Trends and Their Signals

Documented observational evidence is indispensable for translating theoretical climate science into discernible, real-world impacts. Measurements from meteorology, glaciology, oceanography, and terrestrial ecology collectively reveal a coherent signal of planetary warming. This signal manifests in altered synoptic circulation, accelerated polar ice mass loss, ocean thermal expansion, and species range shifts. By synthesising these long-term, high-resolution data records, researchers discern emerging, cross-disciplinary patterns that validate the expectation of anthropogenic climate perturbation and guide ongoing adaptation and mitigation strategies.

A primary strand of observational evidence centres on the rise in global surface temperatures. Long-term surface thermometer readings, satellite datasets, and proxy records, such as ice core isotopic data and tree ring chronologies, yield a coherent temporal picture of thermal variation. The current scientific agreement is categorical: the global mean surface temperature is climbing at a rate that is both rapid and atypical, with the most recent several decades witnessing records that exceed the warmer intervals of the preceding millennia.

The observational framework also includes quantification of global mean sea-level elevation. Data from coastal tide gauges, satellite-based radar altimetry, and geological sea-level markers collectively reveal a sustained increase. This rise, driven in substantial measure by the cryospheric retreat of ice sheets and the thermal expansion of seawater, endangers deltaic floodplains, coastal megacities, and numerous small-island states.

In addition to the markers of thermal and hydrospheric change, observational datasets document shifts in the frequency and intensity of extreme meteorological events. Lengthening heat-wave durations, intensified drought episodes, and a higher incidence of Category 4 and 5 tropical cyclones now exhibit clear statistical signatures that align with the increase in anthropogenic greenhouse gas concentrations. The robustness of these correlations, reinforced by multiple independent lines of empirical observation, signals a future characterised by heightened exposure to such risks.

Evidence collected through observation includes not only visible surface phenomena but also the biological reactions of living systems to altered climates. Ecological variables now document alterations in the distributions, life cycles, and interactions of species within entire ecosystems. For instance, populations of insects, birds, and mammals are consistently recording movements poleward and to higher elevations in search of climates that meet their metabolic and reproductive needs, as earlier preferred zones are rendered unsuitable by shifts in mean temperatures and altered precipitation regimes.

When consolidated, these independent yet convergent biological and physical records form a convincing composite that documents the unfolding of climate change. The patterns laid bare by such scrutiny illuminate the specific, human-forced drivers behind broader planetary warming. Understanding these drivers and their consequences is therefore necessary not only for anticipating future changes but also for engineering timely and equitable responses to both mitigation and adaptation.

Feedback Loops and Climate Sensitivity

Feedback processes are vital to the regulation of the Earth's climate system, directly influencing the degree of climate sensitivity and the preservation of dynamical equilibrium. A feedback loop materialises when a perturbation in one component of the system propagates, inducing a modification that either heightens or mitigates the initial disturbance, thereby provoking an additional response that propagates the cycle.

The ice-albedo mechanism epitomises such a loop. It pivots on the differential reflectivity of ice and the darker substrates revealed beneath. Elevated temperatures diminish the areal extent of continental ice sheets and surface sea ice, thereby exposing seas and tundra that reflect less sunlight. The increment in absorbed solar radiation raises temperatures further, provoking additional ice retreat in a continuously reinforcing progression that influences global radiative balance.

A related but intrinsically different imbalance originates in the thawing of high-latitude permafrost and the destabilisation of methane hydrate reservoirs. Increasing Arctic temperatures promote the microbiological decay of frozen organic substrates, liberating both carbon dioxide and methane, gases characterised by their elevated effective radiative forcing. The resultant atmospheric burden of these gases magnifies the initial thermal forcing, thereby enacting a loop that materially accelerates the pace of anthropogenic climate transformation.

Grasping the intricate network of feedback loops is essential for accurately estimating climate sensitivity—the extent of temperature rise per unit of external perturbation. Climate sensitivity reveals how the planet will react to elevated greenhouse gas levels and thereby governs the severity of forthcoming climate impacts. Incorporating feedback processes into climate models enables researchers to refine their projections and tighten the range of warming anticipated under various emission pathways.

Additionally, pinpointing possible thresholds and tipping points embedded in these feedback loops is critical for forecasting sudden, potentially irreversible shifts in the climate system. Such thresholds might activate cascading processes, precipitating swift and wide-ranging transformations in global climate regimes, biophysical systems, and human communities. Formulating effective climate mitigation strategies and advancing adaptation measures hinge upon a robust grasp of these tipping points and the feedback dynamics that precede them.

Ultimately, feedback loops represent fundamental drivers of the planet's climate behaviour, dictating the system's responsiveness to perturbations and the pathway of future change. Acknowledging and analysing these complex interactions is therefore a prerequisite for sound governance, equitable policy design, and resilient climate action across local, national, and global arenas.

Potential Thresholds and Tipping Points

Thresholds and tipping points are essential frameworks for recognising possibilities of abrupt and irreversible changes in the climate system. When such limits are breached, the resulting cascades can impose widespread, long-lasting consequences for biophysical environments, human societies, and global economies. A systematic investigation of these thresholds is indispensable as uncertainty deepens, driving the strategic framing of mitigation and adaptation. Chief among these concerns is the continuing retreat of polar ice sheets and alpine glaciers, which diminishes planetary albedo and potentially triggers a self-reinforcing cycle of enhanced absorption, accelerated loss, and accelerated sea-level rise. Concurrently, mounting paleo and contemporary observations suggest the Atlantic Meridional Overturning Circulation may be approaching a regime shift. This event would recalibrate precipitation, heat distribution, and storm tracks across northern continents. A third, equally alarming threshold is the warming-induced destabilisation of methane hydrates sequestered in permafrost and under oceanic shelves. The sudden release of this greenhouse gas would impose a powerful and rapid warming feedback. In terrestrial biomes, the risk of crossing thresholds is equally pronounced: mass die-off of Amazon and boreal forests, driven by enhanced drought and fire, would not only convert these regions from carbon sinks to sources, but also reconfigure large-scale biogeophysical properties such as moisture recycling and albedo.

Ocean acidification, driven by the uptake of anthropogenic CO_2, poses an escalating threat to marine biota, potentially culminating in catastrophic shifts across the zooplankton-phytoplankton-benthic pathways that underpin global fisheries and wider biogeochemical cycles. Quantifying the precise thresholds that could precipitate these nonlinear responses necessitates an interdisciplinary amalgamation of contemporary findings in climatology, carbonate chemistry, microbial oceanography, and sedimentological paleorecords. Mitigating against such outcomes, therefore, obliges a forward-looking response that interweaves precautionary regulatory frameworks, emergent biogeochemical engineering, and coordinated multilateral scientific diplomacy. A candid recognition of these critical ecological thresholds can galvanise the formulation of anticipatory, adaptable governance, steering the planet toward a durable, low-carbon future. As understanding of climate-sensitive marine processes matures, differentiated adaptation pathways can be delineated that limit the likelihood of crossing biogeochemical tipping points.

Scientific Consensus and Ongoing Research

Broad agreement on anthropogenic climate change has crystallised only after decades of disciplined inquiry, meticulous observation, and transdisciplinary cooperation. A robust, preponderant community of climate scientists now concurs that the combustion of hydrocarbon fuels and widespread land-use change intensify the radiative forcing of the climate system. This verdict is undergirded by congruent evidence from marine and continental paleorecords, high-resolution satellite and in situ observational networks, and ensembles of multi-decade, high-resolution general circulation and Earth system models.

The Intergovernmental Panel on Climate Change (IPCC), created under the aegis of the United Nations, exemplifies multilateral scholarship in the discipline of climate science. By assimilating the most current peer-reviewed literature and synthesising it into iterative assessment reports, the

IPCC supplies a rigorous evidential foundation for deliberations in international fora. Each cycle of analysis reaffirms the accelerating consequences of human-induced greenhouse gas accumulation and reasserts the critical window for implementing robust, evidence-based climate-risk reduction measures.

Expanded observational capacity and methodological innovation continue to refine the science of climate dynamics. Spaceborne remote-sensing platforms, high-resolution Earth-system modelling, and paleoclimate reconstructions jointly permit researchers to resolve feedback loops and tipping elements at scales of space and time previously unattainable. The iterative output of these investigations underpins ever-more confident probabilistic estimations of future climate pathways, thereby informing risk assessment protocols and adaptive governance.

The inherently polycentric character of climate change obliges researchers to transcend discipline-specific silos. Climate scientists customarily liaise with ecologists, economists, epidemiologists, and engineers to interrogate the conjoined ramifications of altered radiative forcing on natural and anthropogenic systems. Such cooperative scholarship enriches the evidential base for integrated assessment modelling and reinforces the design of measures that anticipate vulnerability and promote resilience across sectors and scales.

To summarise, the prevailing scientific agreement regarding climate change highlights the urgent necessity of coordinated international efforts to reduce its harmful consequences. Sustained research programmes are further clarifying the intricate dynamics of the planetary climate system. They are concurrently developing pragmatic strategies for achieving a sustainable and resilient relationship between humanity and the natural world.

3
Precarious Balance
The Imperiled Biodiversity of America

Biodiversity—the astonishing array of living organisms that inhabits Earth—is a core ecological legacy that cannot be substituted. In the United States, this ecological wealth faces mounting peril from a suite of anthropogenic pressures: the alteration of land, the toxic by-products of production and consumption, the changing climate, and the introduction of non-native species.

When biodiversity is invoked, the reference is not merely the marquee species—like the bald eagle and the grizzly bear—but also the less conspicuous strata of life: the flowering plants, the insects, the fungi, the microorganisms, and the circuitry of interactions that bind them into functioning ecosystems. The resilience and stability of the biosphere depend on this web of interdependencies. Yet the web is fraying, and the risk now is to the fundamental ecological infrastructure upon which all life, including human life, ultimately depends.

Shell remnants of the original, the Earth-spilling green, habitat destruction hurls the United States toward a biodiversity emergency. The relentless sprawl of cities, the intensified monocultures of commercial agriculture, and the unsated subsurface appetite for minerals and fossil fuels are razing, draining, and sealing the land. The immediate result is the slow-motion dissolution of biomes; mountains are carved, marshes are filled, and corridors are severed, generating isolettes that advance the extinction clock for species that require vast, unbroken environments.

Fragmentation of habitats diminishes the amount of living space available to species while simultaneously dividing populations, which hampers interbreeding and the flow of genetic variation. Genetic isolation then increases the likelihood of inbreeding, lowers individual fitness, and raises susceptibility to shifting environmental conditions and disease. The processes that sustain ecological integrity—pollination, seed dispersal, and nutrient cycling—are equally compromised by habitat loss, undermining the resilience and functionality of entire ecosystems.

Pollution compounds these threats, entering environments with stealth and broad reach. Chemicals released by industrial and agricultural sectors permeate ecosystems, contaminating organisms and upsetting critical ecological interactions. Oil spills, airborne particulates, contaminated waterways, and persistent plastics exert cumulative and long-lasting effects that extend far beyond the source.

Each pollution pathway exerts distinctive pressures. Excess nitrogen and phosphorus, originating from fertilised landscapes and urban runoff, trigger harmful algal blooms and hypoxic zones that decimate fish and other aquatic life. Systemic applications of pesticides and herbicides—often beyond the target crop—inflict immediate mortality and insidious sub-lethal changes in species as diverse as insects, birds, and non-target aquatic organisms, thereby unravelling food webs and contributing to population declines.

Human-mediated climate change amplifies the threats already confronting biodiversity. Global warming is restructuring ecosystems and shifting species ranges. The tempo of these alterations frequently exceeds the ability of many organisms to acclimatise or migrate, which raises the spectre of local extirpations and elevated extinction risk. The unevenness of climate change effects, where already imperilled taxa suffer the steepest losses, compounds the complexity of the biodiversity emergency.

Perhaps the most far-reaching consequence of a warming climate is the renegotiation of species' geographic ranges. Widespread thermal increases compel organisms to track isotherms or seek higher elevations, often leading to novel or destabilised community assemblages. Such migratory pressures can collide with human land use, infrastructure, and conservation reserves. Meanwhile, the decoupling of biotic interactions—e.g., altered phenological synchrony in flowering, breeding, or migration—jeopardises mutualisms and trophic dynamics, magnifying extinction cascades among interdependent species.

Alongside climate change, invasive species represent an insidious threat to America's native biodiversity, often lurking unnoticed until damage becomes irreparable. Accidentally introduced or released for horticulture, fishing, or agriculture, these organisms rapidly capitalise on competitive advantages unavailable to local species, monopolising space, food, and reproductive opportunities. They may also introduce pathogens to which native taxa have no resistance. The cumulative effect is the slow, sure unraveling of the complex, co-evolved structures that define America's diverse communities of plants, animals, and microbes; the extinction of a single keystone species can, in a generation, convert a resilient swamp, meadow, or canyon into a simplified, impoverished zone, bereft of the cultural and spiritual landscapes that these organisms once embodied.

Global trade, passenger movement, and climate warming collude to expose new environments to non-native species. Cargo ships, planes, and garden centres supply the seeds; warming winters, excess nitrogen, and drought open the gates. Once in, they use vigorous growth, allelopathy, or novel diseases to transform the rules of a region. Managers, constrained by lagging funding and public disengagement, confront escalating infestations that mock preventative budgets; eradicating a wide-ranging lionfish population or a soil-borne pathogen in an understory demands years of monitoring, herbicides, or reintroducing costly biocontrol agents, often with uncertain outcomes.

Preventing a further, irrevocable loss of native species necessitates a coordinated national mobilisation. Ecological restoration that re-establishes hydrology, native seed banks, and predator-prey links must expand; wildlife corridors must be strengthened to allow native taxa safe passage across increasingly fragmented landscapes. Concurrently, regulations that limit land conversion, filter excess nutrients, and strictly regulate species importation and release must be drafted, enforced, and periodically updated. Only an integrated strategy, spanning local stewardship to international treaties, can blunt the advance of invasive species and sustain the complex, irreplaceable networks upon which both nature and culture in America depend.

Effectively addressing the biodiversity crisis in the United States requires a coherent and collective stance among federal, state, and tribal agencies, conservation organisations, and the private sector. Their shared mandate is to construct and implement forward-looking conservation measures that safeguard entire ecological networks, not merely endangered taxa, thus securing the ongoing viability and adaptive capacity of the nation's natural systems.

America's decline in biological variety offers a vivid illustration of life's interdependence and a compelling ethical invitation to protect the ecological inheritance we pass to the future. Suppose we dedicate ourselves with rigour and consistency. In that case, we can yet foster the recovery and sustained equilibrium of the nation's living systems, ensuring that the rich mosaic of plants, animals, and microorganisms continues to enrich the continent and its people.

Conservation and ecological recovery are imperative, not merely for the worth of nature in its own right, but for the many services that biodiversity delivers to human communities. Pollination of crops, filtration of drinking water, and regulation of atmospheric gases are among the life-supporting processes carried out by diverse organisms. In addition, the broad genetic reservoir contained in native species offers vital ingredients for crops that can withstand climate extremes, for pharmaceuticals that treat disease, and for technologies that mitigate environmental change.

Protecting biodiversity most effectively hinges on the thoughtful creation and stewardship of protected areas. National parks, wildlife refuges, and marine sanctuaries together function as critical safe havens where numerous species may persist without human interference. By enlarging and linking these zones, we strengthen the resilience of entire ecosystems and offer sanctuary to species confronting escalating pressures.

Successful biodiversity conservation, however, demands more than boundary-drawing; it also needs a broad, inclusive framework that respects and incorporates the rights and needs of indigenous peoples and local communities. Indigenous groups, grounded in ancestral knowledge and a deep attachment to their territories, are indispensable to the task of conserving biological and ecological integrity. When their voices and stewardship practices are woven into conservation programmes, their intimate understanding of local dynamics reinforces and enriches our shared endeavour towards protecting the living world.

Education and public awareness stand as vital, yet often overlooked, supports in the larger structure of biodiversity conservation. By embedding environmental literacy in curricula and public outreach, we can nurture a shared appreciation of living systems, transforming passive awareness into active guardianship. When communities understand the interdependence of species and ecosystems, protection evolves from obligation to identity, ensuring that stewardship is not a temporary campaign but a lasting societal norm.

Innovation and technology present equally potent, though distinct, instruments in the conservation toolbox. Remote sensors, high-resolution satellite imagery, and genomics are now more accessible than ever, enabling precise species enumeration, habitat mapping, and genomic rescue, among other strategies. When harnessed under a framework of transparency, equity, and precaution, such tools can translate complex ecological data into actionable conservation insights and cost-efficient interventions, addressing the multifaceted threats to biodiversity in America.

Together, these dimensions form an integrated approach to biodiversity that is not only pragmatic but also ethical. Our commitment must therefore be both collective and persistent, reflecting a deep-seated recognition that biodiversity is a form of public and planetary wealth. By cherishing and legally safeguarding this wealth, we create the ecological, social, and economic scaffolding for a future in which America's rich and dynamic assemblages of life not only survive but thrive, painting a more resilient and prosperous mosaic for every generation that follows.

4
Vulnerable Populations
Facing Disproportionate Risks

Vulnerable populations are diverse groups that face heightened exposure to climate change hazards and diminished ability to cope with those hazards. Their elevated vulnerability arises from a combination of social, economic, and environmental vulnerability, rather than from any inherent trait. Critical to this definition is the recognition that power relations and the historical accumulation of disadvantages, including poverty, discrimination, and marginalisation, shape vulnerability.

Analysing vulnerability, therefore, demands an integrated approach that illuminates the ways income, educational attainment, geographic context, and healthcare access interact. For example, low-income neighbourhoods frequently intersect with flood-prone areas and face inadequate public health infrastructure, compounding risk.

Mapping these intersecting variables permits identification of the populations most likely to suffer amplified harms and hence most in need of targeted support. Parsing vulnerability in this way reveals not only where risk is concentrated but why certain social groups confront disproportionate exposure and harm. Such an understanding is essential for crafting adaptive policies that confront the structural inequities underpinning risk and that enhance the adaptive capacity of the most affected communities. Recognising these connections reinforces the imperative to address climate change not solely as an environmental crisis but as a critical social justice issue.

The notion of vulnerable populations must be understood as transcending traditional demographic classifications and as traversing the layered encounters of intersecting disadvantages. Marginalised groups—whether low-income families, older adults, children, Indigenous nations, or persons with disabilities—confront heightened exposure and reduced adaptive capacity whenever environmental stressors coincide. Long-standing practices of official neglect, policy exclusion, and systemic inequity have enshrined the uneven distribution of ecological burdens and amenities, magnifying the precarity of these constituencies. Attentiveness to the intersecting determinants of vulnerability permits the design of holistic interventions that respond to the distinct and overlapping needs of heterogeneous social segments. Such recognition allows policy and practice to cultivate a genuinely inclusive and justice-oriented framework for shielding the populations most exposed to the ravages of climate change.

Understanding vulnerable populations is not a simple task. It requires a thorough examination of the various factors that contribute to their susceptibility to climate impacts. By dissecting the multi-dimensionality of vulnerability and the structured forms of inequality that underpin it, we lay a solid intellectual and ethical foundation for developing policies that address the specific and compounded hardships experienced by these communities.

Defining Vulnerability in the Context of Climate Change

Vulnerability to climate change is a complex, layered phenomenon combining exposure, sensitivity, and adaptive capacity. Within this context, it denotes the extent to which individuals, households, regions, and ecological systems can be adversely affected by climate stressors, including intensified storms, rising sea levels, and changing temperature regimes. A comprehensive understanding of vulnerability necessitates an analysis of the interlocking social, economic, and biophysical systems that influence a given group's ability to anticipate, withstand, and rebound from climate-related shocks.

Physical location and hazard exposure are necessary but insufficient criteria; the degree of vulnerability is more decisively shaped by the distribution of power, wealth, information, and institutional support among groups. Vulnerability analysis thus serves a critical policy function, guiding the identification of populations and systems that require prioritised attention and the specification of context-sensitive corrective measures. Root-cause identification and the integration of local knowledge are vital to crafting interventions that strengthen adaptive capacity without inadvertently reinforcing existing inequalities. Finally, sensitivity in framing vulnerability is essential to protect the dignity of affected communities; language and practice must empower rather than fixate attention on deficit, ensuring that analysis is always a step toward equitable and inclusive climate governance.

Developing an inclusive and participatory strategy that centres the experiences of those directly affected is crucial for designing equitable and durable responses to climate change. A comprehensive, intersectional framework is necessary to understand how different social groups confront and adapt to climatic stressors. By systematically unpacking the factors that heighten exposure and limit adaptive capacity, we deepen our understanding of the structural inequalities and historical injustices that confront marginalised communities, enabling the formulation of evidence-based interventions that can truly address the roots of vulnerability and drive transformative reform.

Socioeconomic conditions play a significant role in shaping vulnerability under climate change. The interplay of income inequality, resource distribution, educational attainment, and labour market access determines how effectively populations can anticipate, absorb, and recover from climate-related disturbances. Poorer households often face heightened exposure because they lack the financial resources to invest in resilient infrastructure, emergency supplies, or rebuilding efforts after extreme events. Their limited income also leads to deferred or inadequate healthcare, which exacerbates the health threats posed by heat waves, vector-borne diseases, and mental health distress following disasters.

Moreover, historical patterns of systemic discrimination mean that racial and ethnic minorities, older adults, and people with disabilities not only face contextual health and safety disparities but also find that response mechanisms, like evacuation or emergency services, are not tailored to their specific needs. The interplay is compounded by substandard, insecure housing and infrastructure, especially in flood plains—and in areas where policies are discretionary or absent. Such built environments are inherently unable to withstand shocks, trapping low-income residents in a cycle of exposure and limited escape.

Ongoing climate change reinforces the necessity of tackling the root socioeconomic inequalities that shape community resilience and equitable protection. Thoughtfully designed policies, collaborative grassroots projects, and strategically resourced assistance programmes can reduce the influence of these disparities on exposure and sensitivity, thereby promoting inclusive, just outcomes in adaptation and emergency interventions.

Health Impacts on Marginalised Communities

Marginalised communities experience climate change in ways that compound existing inequities, translating into pronounced health impacts. Limited access to preventive care, fragile infrastructure, and adverse socioeconomic conditions heighten their vulnerability to both physical and mental health disturbances. Intensified air pollution translates into higher rates of asthma and other respiratory ailments, while the combination of soaring temperatures, substandard housing, and urban heat islands escalates the prevalence of heat-related illnesses.

Equally alarming are the psychological sequels: economic disruption, forced migration, and trauma linked to extreme weather events, which engender chronic stress and mental health degradation. The ripples of these health burdens extend from individuals to families, neighbourhoods, and the broader social fabric, destabilising entire communities.

CLIMATE CHANGE IN AMERICA 53

Addressing these disparities is therefore a matter of justice and public health. Effective responses must include accessible and culturally competent healthcare, the reinforcement of social and economic safety nets, and the promotion of community-led decision-making. Incorporating Indigenous knowledge and traditional ecological practices into these frameworks not only respects cultural sovereignty but also reinforces a holistic view of health that recognises the inseparability of people and their environment. Such integrative strategies are essential for fostering durable resilience and securing equitable health outcomes amid a changing climate.

Sustained collaboration among policymakers, public health professionals, and community representatives is essential to address the specific vulnerabilities faced by marginalised populations. Such an approach ensures that interventions are equitable, culturally relevant, and strategically aligned with community capacities. By centring disadvantaged groups within planning, implementation, and evaluation phases, we can work towards a society where environmental hazards are met with protective infrastructure, accessible health services, and economic opportunities, allowing every person—irrespective of income, race, or geography—to prosper within a climate-resilient future.

Indigenous communities, historically tethered to their ancestral territories, now confront environmental injustices that climate change magnifies. Generations of marginalisation have predisposed these populations to a cascade of ecological hazards, the severity of which outstrips their numerical proportion, leaving them with the least adaptive capacity. This chapter investigates their distinctive vulnerabilities, resilience practices, and the intersection between historical disenfranchisement and contemporary environmental decision-making. Safeguards that indigenous peoples historically engaged—valorising soil, water, and biodiversity—now confront accretive climate stresses and resource exploitation. Their intergenerational ecological knowledge constitutes a repository of adaptive strategies yet remains under-leveraged owing to limited political visibility and resource deprivation. When climate extremes, invasive species, and altered seasonal cycles disrupt hunting, fishing, and gathering, the consequences extend beyond economics, eroding cultural identity tied to biophysical cycles. Simultaneously, encroachments for mining, hydropower, and industrial agriculture, frequently sanctioned without free, prior, and informed consent, fracture territorial integrity and adaptive capacity. Remedial action must then centre on co-governance that respects territorial sovereignty, integrates indigenous worldviews into scientific models, and operationalises the United Nations Declaration on the Rights of Indigenous Peoples across climate and environmental policies.

Collaborative models that enable Indigenous communities to engage fully in decision-making and to jointly conceive effective remedies are central to realising environmental justice. Elevating Indigenous voices, respecting ancestral knowledge, and embedding Indigenous perspectives in adaptation and mitigation efforts are non-negotiable steps. In addition, cultivating symbiotic relationships between Indigenous nations and governmental bodies can generate responses that are fair, sustainable, and culturally anchored.

The decisive move towards a more equitable and resilient future requires a concerted effort to recognise and dismantle the entrenched barriers that perpetuate environmental injustice against Indigenous peoples.

Urban vs. Rural Challenges

In the unfolding climate crisis, cities and countryside alike must navigate distinct vulnerabilities that require nuanced and situational responses. Metropolitan areas routinely encounter demands that outstrip existing infrastructure, elevated concentrations of airborne contaminants, intensified temperatures due to the heat island phenomenon, and a shortage of parks and tree cover. Each of these stressors is magnified by climate change, threatening the fragile equilibrium that sustains urban life. Conversely, the more diffuse settlements of rural regions experience their form of exclusion: sparse healthcare networks, under-resourced emergency services, and a dependence on weather-sensitive livelihoods that leave residents perilously exposed to droughts, floods, and shifting growing seasons. The irrefutable divergence in exposure and capacity to respond mandates that climate-resilience planning be both disaggregated and inclusive, ensuring that no community is left to face the crisis with disproportionate disadvantage.

Cities concentrate human activity and thus economies of scale in both production and social interchange. Yet, the same concentration that fuels dynamism also magnifies vulnerability. The acceleration of urban populations has outpaced the retrofitting of physical systems, generating higher greenhouse gas discharges and exhausting freshwater supplies. In turn, the surge of airborne toxins and solid waste constitutes an ongoing public health emergency, with the sickest burdens falling on low-income neighbourhoods already overburdened by structural inequities. The geographical dominance of impermeable substrates, when coupled with the chronic shortage of trees and parks, intensifies the heat island phenomenon, compounding the incidence of heat-related illness and mortality, especially among the elderly and the very young. For both contexts, the separation between exposure and adaptive capacity underlines the necessity of targeted intervention calibrated to specific demographic and geographic contingencies.

Conversely, rural areas face a distinct array of challenges. Sparse proximity to health clinics and emergency services magnifies their susceptibility to disasters intensified by climate change. Agricultural districts are further destabilised by shifting rainfall regimes and elevated temperatures, which jeopardise yields and threaten food sovereignty. These climatic strains are intensified by socio-economic pressures, including shrinking job prospects and the out-migration of younger residents, which entrench poverty and widen inequalities within the countryside.

To meet the contrasting needs of urban and rural constituencies, policies must be precise and attuned to local realities. Urban resilience frameworks ought to direct funds toward green roofs, expanded transit, and inclusive land-use policies that reduce flooding and heat. At the same time, rural programming must elevate transport networks, expand clinics, and adopt climate-smart farming. Bridging the urban–rural fault line is also essential, since the vitality of cities often rests on the well-being of hinterlands, and joint problem-solving can yield more durable outcomes.

A clear-eyed appraisal of urban and rural vulnerabilities allows for the design of holistic responses that protect every citizen, ensuring climate preparedness that is both equitable and enduring.

Case Studies: Human Faces of Risk

To grasp the full implications of climate change for society's most vulnerable, we must confront detailed, on-the-ground accounts that translate broad climate data into lived experience. Such accounts render the statistics personal, illuminating discomforts that policy change alone cannot relieve. One illustrative example emerges from Louisiana's vulnerable coastal wetland fringes, where coastal First Nations maintain a cultural mosaic intimately braided with marshes, bayous, and oyster beds. Accelerated sea-level rise, compounded by stronger and more frequent hurricanes, is outpacing community-directed levee projects and eroding deltaic soils. As marshes retreat, cemeteries and subsistence plots are subsumed by saline water, undercutting both a braided economic and a spiritual livelihood.

Further inland, the Central Plains present another vignette

of exposure. Generational grain and livestock farms, already rote with modest margins, are experiencing a departure from agronomic probabilistic norms. Multi-year droughts now purchase co-insurance with unprecedented high-temperature grain fill periods, while restive flood events deliver delayed planting windows and topsoil siltation. Such volatility is pressing rural mental health resources already stretched by long-standing economic marginality, revealing that crops are vectors not only of protein but of social distress, as family stress, suicide rates, and hypertension convergence demonstrate. Together, such micro-narratives weave a larger indictment of adaptation challenges that are not strictly environmental but social, cultural, and psychological.

Moreover, in metropolitan areas like New York City, low-income neighbourhoods endure the longest exposures to extreme heat and elevated air pollution, compounding chronic health conditions among residents who typically have reduced access to both healthcare services and public green spaces. Such patterns illustrate the layered and systemic nature of climate hazards, underscoring the pressing requirement for interventions specifically designed to address the needs of the most affected populations.

Accounts from these neighbourhoods reveal the inseparable linkage among socioeconomic disadvantage, environmental vulnerability, and health inequity, demonstrating that effective policy must integrate all three dimensions. Engaging with these testimonies enables researchers and decision-makers to connect empirical data with lived experience, cultivating the empathy and insight necessary to advocate for proactive, equitable climate strategies that protect the people who are already bearing the heaviest burden.

Policy Frameworks and Support Mechanisms

To confront the climate-change-related challenges confronting the most vulnerable populations, it is essential to build sound policy frameworks and accompanying support mechanisms. These instruments guarantee that the distinctive needs of at-risk communities are fully recognised and that their representatives are integral to the deliberative process. A sound framework is centred on the creation and execution of policies that directly target the specific vulnerabilities of marginalised populations. Completing this task requires systematic risk profiling, participatory risk-assessment exercises, and the ongoing incorporation of community members at every stage of policy conception, modification, and enforcement. Support mechanisms must then concentrate on facilitating equitable access to financial, technical, and informational resources, empowering communities to develop adaptive capacities and long-term resilience to changing climate realities. Collaboration between state institutions, civil society organisations, and local community leaders is indispensable. Such a multi-stakeholder model ensures that varied experiences and expertise are brought to bear, resulting in policies that are both comprehensive and contextually relevant. Furthermore, policy frameworks must explicitly embed equity principles, designing interventions that remediate pre-existing inequalities and guard against the further marginalisation of already vulnerable groups.

Prioritising equity in the formulation of policies enables leaders to begin the long-overdue work of remedying past injustices while securing equitable results for everyone in the present and future. Equally important is the systematic development of monitoring and evaluation frameworks that gauge the impact of every adopted policy and support measure. Ongoing analysis of both successes and shortcomings provides the empirical basis for timely revisions, ensuring that each intervention adapts to shifting circumstances and persistent vulnerabilities. Equally, investment in the capacities of marginalised communities is necessary.

When residents and local organisations acquire the information and expertise to influence policy dialogues, they fortify the long-term viability of every measure and embrace genuine self-governance. Transparent and accountable institutional designs further promote trust, inviting affected groups to engage in consultations that shape policies. International cooperation, finally, is indispensable for managing the far-reaching and border-crossing consequences of climate change on at-risk populations. Through collaborative arrangements, states and organisations can share finances, distribute knowledge, and implement synchronised strategies, enhancing fairness and effectiveness. In combination, comprehensive policy ecosystems and responsive support instruments constitute fundamental means of protecting the welfare of populations that climate change jeopardises, while the climate adaptations themselves are still in formulation and execution.

The Role of Community Engagement and Advocacy

Community engagement and advocacy constitute vital, interrelated strategies for mitigating the disproportionate impacts of climate change on vulnerable populations. As communities confront the entangled ecological, economic, and social dimensions of their entitlements, the necessity of affording them institutional avenues for substantive engagement and protective advocacy becomes evident.

Central to community engagement is the systematic elevation of indigenous technical knowledge and lived experience. This process entails structuring forums through which residents articulate their perceptions of climate hazards, their adaptive aspirations, and their immediate resource needs. When the lived realities of marginalised constituencies are honoured and integrated into governance cycles, policymakers and external partners receive essential, context-specific intelligence regarding the intersections of vulnerability. Such an inclusive governance modality not only enhances the situational analysis available to decision-makers but also cultivates communal ownership of policy and programme vectors, thereby enabling collective agency in the design and enactment of adaptive interventions that are congruent with the community's cultural and material conditions.

Effective advocacy is forged through deep community engagement, ensuring that those experiencing the effects of climate change occupy the centre of decision-making. Targeted advocacy campaigns designed by community leaders can illuminate the acute imbalances that climate-related hazards impose on the most vulnerable. Such visibility becomes the impetus for galvanising both local partners and a wider constellation of stakeholders into coordinated, sustained action.

Equally important, community-centred advocacy retains the power to hold public authorities accountable for systemic inequities that deepen exposure and risk. By uniting around policies that mandate fair distribution of resources, equitable healthcare access, and climate-resilient infrastructure, communities can sculpt policies that confront vulnerabilities at their origin, embedding equity into the very frameworks that govern recovery and prevention.

Yet, the interplay of advocacy and engagement reaches far beyond attending rallies or public hearings. It is sustained through capacity-building that empowers residents with the analytical and practical skills required for both immediate adaptation and the long-term mitigation of climate risks. By prioritising education, vocational training, and the co-design of innovative, locally rooted adaptation measures, communities can sovereignly steer their trajectory toward lasting resilience.

At the heart of successful community engagement and advocacy lies the systematic development of intersectoral partnerships and collaborative networks. When municipalities, governmental bodies, non-profit agencies, academic consortia, and private sector actors converge, they can pool financial, intellectual, and technical assets, permitting a holistic and multidimensional response to the full spectrum of climate vulnerabilities.

In this context, community engagement and advocacy are not peripheral actions but, rather, decisive mechanisms of structural transformation. They embed the principles of resilience and sustainability within the political and social fabric of society. Confronting the expanding complexities of climate change, the capacity of communities to organise, influence, and act emerges as a non-negotiable pillar of a global response that is both protective of and empowering for the most at-risk populations.

Concluding Thoughts: Pathways to Resilience

Responding to current climate pressures demands that we prioritise resilience in marginalised populations. Success hinges on strategies that recognise the intertwined social, economic, and environmental factors affecting lives. A cornerstone of any effective resilience effort is the cultivation of community ownership and empowerment. Support must be directed toward enabling residents to engage meaningfully in the design and execution of policies that shape their futures. Elevating local voices in forums of influence not only democratises the process but also enriches it, lending legitimacy to strategies that seek both equity and sustainability. In parallel, the careful integration of indigenous knowledges and local customs can supply tested strategies for weathering ecological shifts. Durable partnerships among government, civil society, and grassroots organisations are therefore non-negotiable, ensuring that intervention efforts are coherent and culturally attuned. Resources devoted to education and vocational training amplify a community's capacity to respond, creating a multiplier effect whereby individuals transform vulnerabilities into advocacy and adaptation.

Addressing systemic inequalities that exacerbate vulnerability constitutes a necessary first step. This requires not only dismantling institutional barriers, but also rectifying the legacy of historical injustices that have concentrated danger in marginalised communities. Agencies and institutions must orient resource allocation toward equity, ensuring that every community acquires fair access to essential support structures. Targeted interventions and anticipatory policy design can therefore attenuate the unequal effects of climate change on those already exposed to intersecting risks. Concurrently, fostering sustainable economic pathways and diversifying livelihoods will deepen resilience. Strengthened economic foundations, combined with inventive, ecologically sound practices, can fortify communities against future shocks while promoting long-term adaptive capacity.

Preserving and reinforcing cultural resilience is equally essential for protecting the distinct identities and traditions of indigenous nations. Valuing and integrating traditional ecological knowledge deepens our collective repertoire of sustainable management practices. A sincere acknowledgement of past injustices, paired with a commitment to fostering indigenous guardianship of the land, is vital to constructive resilience. By centring collaborative processes that respect ancestral connections and cultural heritage, we can nurture adaptive strategies that also honour the spiritual and historical dimensions of communities, thus advancing a more balanced and respectful coexistence with the Earth.

Though obstacles mark the journey to resilience, it remains attainable. When we cultivate a culture of solidarity and collective obligation, we create pathways to solutions that consistently uphold and safeguard the most vulnerable among us. Such progress requires unwavering compassion, determined persistence, and a collective pledge to forge a future that is both fair and inherently resilient for every member of our global community.

5
Regional Impacts
Case Studies Across the Nation

Introduction to Regional Climate Impact Analysis

Regional climate impact analysis is essential for grasping the nuanced, interconnected challenges climate change is imposing on the United States at a sub-national scale. As global warming continues, evidence grows that different areas are the sites of distinctive—and often acute—alterations to the physical climate system. These alterations, in turn, disrupt local ecosystems, strain economic sectors, damage infrastructure, and affect human communities. A thorough understanding of these region-specific pressures is therefore imperative for crafting well-informed, localised responses. By investigating the precise vulnerabilities and resulting challenges that each region confronts, decision-makers can formulate precision-engineered strategies that reduce risk and enhance adaptive capacity. The interplay between biophysical, societal, and economic variables underscores the necessity of integrative analysis if resilience and successful adaptation are to be achieved.

When viewed through the framework of regional climate impact analysis, the effects of climate change emerge with telling specificity. Rising sea levels and intensified coastal erosion are reshaping the Northeastern states, while the Midwest grapples with the dual threats of shifting rainfall patterns and soil degradation that complicate the management of agriculture and water resources. In the South, mounting heatwave frequency and rising cooling demand stress energy grids and public health systems alike. The West confronts successive, more severe wildfire seasons that alter carbon budgets and air quality, while the Pacific Northwest is witnessing phenomena that destabilise salmon lifecycles and forest health. Simultaneously, declining snowpack across the Rockies is already compressing reservoir refill seasons and threatens downstream irrigation. Urban centres throughout the country must urgently retrofit electric grids, drainage systems, and transport networks.

At the same time, Indigenous nations continue to assert the need for protection of ancestral lands and living cultures that are themselves ecosystems. This structured overview illustrates how climate change, while globally uniform in its cause, manifests through regionally layered and distinctive vulnerabilities. The analysis thus primes the next analytical tier: comparative case studies that illuminate localised adaptation pathways and risk mitigation opportunities across the continental U.S.

Northeastern States: Rising Sea Levels and Coastal Erosion

The northeastern United States—comprised of Maine, New Hampshire, Massachusetts, Rhode Island, Connecticut, New York, New Jersey, Pennsylvania, Delaware, Maryland, and the District of Columbia—is confronted by the dual hazards of accelerating sea-level rise and progressive coastal erosion. Concomitant global warming is translating into higher ocean temperatures and the thermal expansion of seawater, producing a markedly steep climb in mean sea level along the northeastern shoreline. This trajectory endangers not only the fragile marsh and estuarine systems that undergird coastal biodiversity, but also the infrastructure and settlements that sustain the region's historic and densely populated urban centres.

Pressures on these coastal economies are immediate and acute: saltwater intrusion into aquifers, contamination of potable water, and diminished shellfish stocks are forcing a reassessment of livelihood strategies for the mill-labour ocean-dependent population. Rising water lines also translate into expanded and intensified flood zones for successive storm surges, extending the inundation hazard far into low-lying interior corridors. Consequently, essential infrastructure—bridges, highways, power grids, and health and safety stations—faces a higher probability of prolonged impairment and catastrophic failure.

In light of these mounting risks, an interdisciplinary response is forming. Coastal engineers are advancing levee and living shoreline designs that temper wave energy while fostering ecological resilience. Urban planners are retrofitting drainage basins, applying green-absorption strategies, and strategically relocating vulnerable public facilities. Meanwhile, state and municipal policymakers are developing regionally coordinated, adaptive land-use regulations that incorporate sea-level trajectory forecasts into zoning and permitting processes. Collectively, these interventions are intended to safeguard both the ecological and social fabric of the northeastern seaboard in an era of climatic unpredictability.

Coastal resilience and adaptation are being strengthened through green infrastructure investments and carefully coordinated land use planning. Equally important is the active partnership of federal, state, and local agencies, whose joint effort is essential to managing the complex and escalating hazards of rising sea levels. Beyond the physical phenomena, northeast rising tides introduce pronounced social and economic pressures. The displacement of communities, the progressive erosion of irreplaceable historic sites, and the jeopardy faced by tourism and fisheries all undermine the region's cultural continuity and economic viability. Confronting these overlapping challenges requires an approach that is holistic and participatory: one that integrates grassroots input, interdisciplinary inquiry, and policy processes that refuse to exclude any affected group. Recognising the mutual dependencies of these domains enables all stakeholders to forge communities that are both resilient to the present and adaptable to future variability.

Midwestern Challenges: Agriculture and Water Resources

The Midwestern United States has long borne the designation of the nation's breadbasket, its broad fields and high-yield farms sustaining vast quantities of grain and livestock. Yet, this essential agricultural base now confronts daunting pressures that arise from a changing climate. Most consequential are the shifting rainfall regimes that result in both concentrated deluges and extended dry spells. Such variability jeopardises the finely tuned calendars of planting and harvest, undermines the expectations of yield, and compromises the structural integrity of soil. Augmenting this predicament are the more frequent, more intense, and less predictable extremes—flash floods, prolonged wind events, and tornadic activity—that further weaken the land.

Simultaneously, the task of stewarding the region's water resources has grown more complicated. Farmers and livestock producers, long accustomed to drawing from both surface reservoirs and the vast regional aquifers, now face the dual spectres of declining reservoir levels and waning aquifer recharge. Each irrigation season deepens concern over the sustainability of the groundwater lifeline. Compounding this picture, the very run-off—now more concentrated and laden with fertilisers and herbicides that accompany heavier rainfall—endangers both potable supplies and the delicate balance of Midwestern ecosystems. Efforts to maintain agricultural viability while safeguarding water quality are thus intertwined, demanding adaptive policies that anticipate further climatic variability.

To meet these challenges, the Midwestern agricultural sector is advancing a suite of innovative strategies designed to enhance resilience. Precision agriculture is at the forefront of these efforts, employing sophisticated sensor and satellite technologies to improve the efficiency of water, fertiliser, and pesticide application while reducing the carbon footprint of farming operations. Complementing these efforts are sustainable water management techniques such as tiered rainwater harvesting systems and reduced-tillage practices, which together buffer the sector against increasingly unpredictable precipitation and bolster long-term water availability.

Successful adaptation hinges on active partnership among farmers, scientists, extension educators, and policymakers. Policymakers can facilitate wide-scale adoption by integrating crop insurance reforms that reward diversification and by supporting research on region-specific cultivars. Simultaneously, upgrading ageing irrigation systems and embedding real-time monitoring tools ensure water is applied only when and where it is needed. Public outreach campaigns that demonstrate water-efficient technologies and promote conservation are equally vital in embedding resilience at the farm level.

Ultimately, embedding climate-smart practices into the region's broader agricultural and water governance framework is essential to protecting the Midwest's keystone contribution to national food security. Through sustained innovation and synergistic collaboration across disciplines, the region is positioning itself to adeptly manage the risks posed by shifting climate regimes while honouring its agricultural heritage.

Southern States: Heatwaves and Energy Demand

Heatwaves in the U.S. South are confronting the region with accelerating risks to health, ecosystems, and energy infrastructure. Temperatures are climbing, and the frequency of severe heat events is increasing, pushing electrical grids and public health systems to their limits. The hazardous combination of high heat and oppressive humidity multiplies heat-related illnesses and drives up cooling loads, creating interlinked challenges that demand multifaceted, long-term planning. Low-income households and the elderly remain the most exposed, as they often lack affordable access to air conditioning, reside in overheated dwellings, and are more susceptible to heat stress. The repercussions transcend human health, impairing crop yields, straining water resources, and disrupting natural habitats. In light of these threats, Southern states are crafting and deploying adaptive measures that cut across energy policy, urban design, and public health. Among the actions being taken are the refinement of heat alert systems, the strategic placement of cooling centres, the enforcement of energy-saving building codes, the promotion of reflective pavements and green roofs, and the acceleration of a diverse, resilient energy mix.

Effective responses to the complexities of extreme heat in Southern states depend on partnerships among public agencies, non-profit organisations, utilities, and local constituencies. Recognising the inseparable links among climate change, rising energy consumption, and population health during heatwaves enables a coordinated strategy that strengthens communities against immediate and prolonged stress. This strategy must include rapid protective actions and forward-looking adaptation and emission-reduction measures that together foster enduring climate resilience. With sustained collaboration and creative policy design, the Southern states can transform the current heatwave threat into a springboard for leadership in climate resilience and sustainable growth.

Western Wildfires: Causes and Consequences

Wildfires in the Western United States are now a pressing reality, endangering human safety, property integrity, and ecological stability. The ignition and spread of these fires arise from overlapping influences. Elevated temperatures, prolonged drought, and erratic precipitation patterns—traits of a changing climate—create an ideal fire environment. Compounding these climatic stresses are land management legacies: decades of aggressive fire suppression, coupled with the gradual buildup of dense, drought-sensitive foliage, have amplified fuel loads. Human agency is still a decisive factor: campfires ignited out of season, a tossed cigarette, or a malfunctioning piece of machinery can convert a vulnerable landscape into a conflagration. Once the first flame takes hold, dry fuels, low humidity, and high winds allow fire tongues to multiply with alarming speed. The aftermath is measured in irreversible loss. Lives vanish, neighbourhoods become ash, and entire watersheds are altered. The charred wood that once sheltered mammals and songbirds now accelerates run-off and exposes soils to erosion, threatening aquatic systems downstream. Smoke plumes cross state lines, seeding compromised lungs and triggering public health emergencies. The fiscal strain is equally stark; suppression campaigns, home reconstruction, and lost economic output amount to the billions. To confront this compound threat, a holistic strategy is imperative. Firebreaks, intentional burns, land-use planning, and community preparedness must intersect with policies that address the climatic changes already in motion.

Enhanced forest stewardship involves implementing controlled burns and selective thinning to mitigate excessive fuel accumulation. Allocating resources to early-warning detection systems and constructing fire-resilient infrastructure strengthens both preparedness and response. Coordinated efforts among federal, state, and local agencies—coupled with community engagement in fire safety training—constitute essential dimensions of an integrated wildfire management framework. Furthermore, advancing empirical research on fire dynamics, climate projections, and ecological restoration enriches policy-making with evidence. Addressing underlying drivers and enacting preventive strategies are critical to attenuating the damaging consequences of wildfires on ecosystems, economies, and public health.

The Pacific Northwest: Marine Ecosystems and Deforestation

The co-occurrence of temperate rainforests, snow-fed river systems, and extensive marine environments characterises the Pacific Northwest of the United States. The stability of this interlinked biogeographic province is increasingly jeopardised, with large-scale forest removal representing an especially acute risk. Accelerated by surging timber markets and the spread of urban footprints, the clearance of old-growth and second-growth stands has fragmented vertebrate and invertebrate corridors, destabilised hydrological cycles, and altered nutrient flux. The resulting loss of forest biomass similarly increases atmospheric carbon concentrations, thereby reinforcing the pernicious feedback loops of climate forcing. Simultaneously, the nearshore and pelagic zones of the Pacific margin confront a triad of stressors. Spanning from kelp-dominated intertidal shelves to deep submarine canyons, these regions serve as critical nurseries for both economically and ecologically crucial taxa. However, globally heightened sea temperatures, the concurrent decline in carbonate saturation, and the introduction of contaminants are eroding this biological wealth. Specifically, declining carbonate saturation is diminishing the calcification rates of pteropods and bivalves, organisms that serve as foundational prey. Coupled with thermal stratification that redistributes productive upwelling, these stressors are redrawing the ecological maps of the region, frequently to the detriment of indigenous and biomass-rich assemblages.

Furthermore, pollution from terrestrial and marine sources continues to intensify the difficulties confronting the marine ecosystems of the Pacific Northwest. The release of plastics, toxic chemicals, and various other contaminants not only inflicts direct damage on marine organisms but also undermines the ecological integrity of the habitats on which they depend. Effectively confronting these interconnected challenges demands an integrated strategy that fuses responsible land-use planning, robust conservation initiatives, and enforceable regulations designed to curtail pollution. Meaningful progress hinges on cooperation among federal, state, and tribal agencies, non-governmental organisations, and coastal communities, all of which must work collaboratively to craft and execute measures that protect these ecosystems while also addressing the parallel threat of deforestation. Elevating public comprehension of the reciprocal dependence between terrestrial and marine realms, and of the irreplaceable value of biodiversity, can cultivate a cooperative ethos committed to preserving the Pacific Northwest's natural heritage for the benefit of generations yet to come.

Mountain Regions: Melting Snowpack and Water Supply

Mountain regions serve as crucial reservoirs for downstream water supply, rendering accelerated snowpack melt under climate change a critical concern. Rising air temperatures have shifted the seasonal snowmelt curve, producing earlier run-off peaks that diminish summer water availability during peak demand months for municipalities, agriculture, and recreation, while simultaneously destabilising ecosystem water balances. These altered hydrological patterns require that researchers and resource managers refine their understanding of the mountain hydrological system. The following discussion synthesises the mechanisms through which snowpack loss is transmitted downstream, surveys adaptive interventions already underway in elevation-dependent water systems. It assesses the implications for broader water governance in the region. The analysis encompasses altered hydrological timing, cascading effects on aquatic biogeochemistry, and the potential for conflicting priorities among users in interconnected watersheds, including those governed by transboundary treaties. Addressing these interconnected challenges demands coordinated, interdisciplinary engagement to devise adaptive governance, infrastructure, and ecosystem-conservation strategies that together sustain water security in warming mountain environments.

Urban Areas: Infrastructure Resilience and Adaptation

City environments are now the frontline of climate change impacts, facing intensified risk owing to high-density living patterns, intricate built environments, and concentrated economic functions. Rising average temperatures and the escalating frequency of extreme events are steadily widening the vulnerability of essential infrastructures. This section emphasises why enhancing and adapting urban infrastructure is critical for sustained urban viability.

Structural resilience in cities includes transport arteries, energy and water networks, digital systems, and the built fabric. Each component must be engineered not only to resist foreseeable shocks but also to recover reliably after disturbances. Coastal cities, for instance, must address encroaching seas and recurrent flooding that compound the risk of landward saltwater intrusion in drinking supplies. Concurrently, prolonged heat and violent storms can overload power grids and transport networks, triggering cascading outages. City resilience planning, therefore, pivots on pre-emptive risk assessment, capital in reinforced design, and the deployment of sensing and flexible technologies that permit real-time mitigation and adaptive capacity.

Effective urban adaptation extends beyond retrofitting physical assets; it necessitates the coevolution of policy frameworks, governance mechanisms, and participatory processes. Planners and officials are charged with restructuring legacy networks—transport, water, energy, and communications—so they can tolerate and recover from a warming climate, while systematically locating and fortifying nodes of highest vulnerability. Equally critical is the deliberate design of adaptation pathways that guarantee marginalised populations gain equitable protection and are not conscripted into compensatory stress.

Global evidence illuminates a diverse menu of adaptation tactics. Cities like Toronto and Melbourne have integrated green roofs, bioswales, and urban forests, employing nature-based systems that moderate both stormwater efflux and the urban heat island effect. Competing cases, such as Rotterdam and New Orleans, have emphasised the fortification—or replacement—of levees, drainage systems, and critical electrical grids against heightened coastal and fluvial extremes. Complementary to these, digital technologies—microgrid controllers, sensor networks, and predictive analytics—are woven into both old and new systems, creating an adaptive capacity to dynamically manage climate hazards.

Achieving the desired level of resilience thus mandates synergistic orchestration of technical design, adaptive legislation, community partnership, and scaled financial underwriting. The resultant integrated model of governance and investment not only protects essential urban services but also forwards the broader goals of environmental stewardship and community vitality within an uncertain climatic future.

Indigenous Lands and Cultural Heritage Preservation

Indigenous territories throughout the United States are imbued with historical, cultural, and ecological significance, forming the bedrock of the nations that have called them home for millennia. These landscapes are more than physical space; they are living archives of traditions, languages, and ceremonies, intimately woven into the identities of the tribes. Yet, their continued vitality is jeopardised by the accelerating effects of climate change. Rising temperatures, shifting precipitation patterns, and an uptick in extreme weather are altering the very ecosystems upon which traditional lifeways depend, threatening cultural practices tied to land and water. Protecting these territories and their accompanying cultural heritage is thus a matter of socioeconomic and environmental justice, since the health of these ecosystems is inseparable from the health of the tribal communities that steward them. Indigenous ecological knowledge, distilled over generations, provides frameworks for resilience that the broader society is beginning to recognise as vital for planetary health. Within this perspective, the Indigenous bond to land is understood as a practical, living blueprint for climate adaptation. This section examines the particular vulnerabilities climate change poses to Indigenous territories, presenting detailed case studies that reveal both exposure and adaptive agency. It also documents synergistic partnerships among tribal nations, federal and state land-management agencies, and conservation organisations, illustrating how collaborative governance can safeguard these irreplaceable landscapes for future generations.

In addition, this chapter will explore how the preservation of cultural heritage and climate action can meaningfully converge, highlighting the critical need to centre Indigenous sovereignty in environmental policy and decision-making processes. By collaborating with Indigenous communities and honouring their rights, policies can be crafted that effectively protect both the ecosystems and the cultural legacies embedded within them. This section clarifies how Indigenous territories, cultural practices, and adaptive capacity are mutually reinforcing, to foster constructive conversation and concerted action that will sustain the integrity of these irreplaceable landscapes for future generations.

Conclusion: Integrated Approaches for Multi-Faceted Threats

Our analysis of the climatic shifts confronting the nation has revealed that the full spectrum of threats cannot be countered in isolation. Protections for Indigenous territories and cultural heritage must therefore be woven into the larger fabric of climate mitigation and adaptation. Through the comparative examination of distinct regional case studies, we have documented the particular exposures that each area confronts, reinforcing the imperative that any response be comprehensive, cross-jurisdictional, and sensitive to the interdependence of natural and cultural systems.

At the heart of crafting coherent climate responses lies the understanding that the enormity of the challenge overwhelms any solitary organisation. Only through sustained collaboration among federal, state, and local authorities, Indigenous nations, non-profit organisations, scholarly institutions, and the private sector can we weave the diverse knowledge and assets needed to untangle climate-related threats. Joint action magnifies our ability to respond to the societal, ecological, and economic dimensions of climate impacts that are so deeply linked to one another.

True coherence, however, requires that we advance strategies on multiple levels: we must weigh ecological realities alongside economic vulnerabilities, cultural values, and equity demands. Solutions that leave social disparities unexamined risk propagating the very inequalities that poverty and environmental decline often magnify. Thus, our planning must be participatory, explicitly inviting those who have been marginalised to guide priorities and strategies, and striving to heal the inequalities that previous policies may have hardened.

Execution of unified strategies must be multi-faceted and future-oriented, tying together mitigation and adaptation. Funding and building green infrastructure, scaling up renewable generation, restoring ecosystems, and harnessing emerging technologies will simultaneously curb emissions and elevate community resilience. Alongside these investments, we must ensure that vulnerable populations have the training, institutional support, and equitable access to knowledge and capital needed to prepare for and recover from climate shocks. Only then will resilience be robust, equitable, and sustainable for generations to come.

A genuinely cohesive framework must therefore be anchored in international partnership and the transnational exchange of expertise. Because the climate respects no frontiers, sound interventions developed in one locality can illuminate pathways elsewhere. Cultivating collaborative ecosystems and nurturing global solidarity not only fortifies our joint resistance to shocks but also magnifies the ethos of common accountability that the planetary emergency demands.

The odyssey toward harmonised solutions for varied trials accordingly requires determination, inventiveness, and an unflagging pledge to preserve the Earth and its populations for those yet to come. By confronting the intricate local consequences of change and grasping the web of global interdependence, we can rally around a shared mission and concerted measures, emboldened by the conviction that united, we can surmount even the gravest of trials.

6
Federal Environmental Policy
A Historical Overview

Origins of Federal Environmental Policy

The beginnings of U.S. federal environmental policy were conceived without an evident appreciation of their ecological longevity. Early policy developments appeared alongside the popularisation of conservation and preservation ideas in the public arena. During the late nineteenth and early twentieth centuries, the pace of industrial expansion and urban sprawl brought about acute air and water contamination, widespread deforestation, and the erosion of finite resources. The dominant perspective, however, was that of prioritising economic growth; policies crafted during this period were oriented toward the short-term maximisation of productive output and wealth, even when this collided with environmental integrity. Nevertheless, public and private advocacy for the protection of specific landscapes, for wildlife security, and planetary health began to take root. The resulting measures of the day were generally enacted at the local and state levels, resulting in a patchwork of prohibitions and incentives. Such fragmentation prompted a consensus that only a coordinated, federal strategy would suffice to meet the sophisticated, interrelated challenges for which the earlier, piecemeal responses were proving inadequate.

It became clear that an organised federal response was required to confront problems that crossed state and regional borders. Although early efforts advanced slowly and lacked a clear national direction, they created a foundation for what would eventually become a cohesive federal environmental policy. Expanding public concern for environmental degradation during this period encouraged a series of landmark legislative actions and, eventually, the establishment of federal agencies and regulatory frameworks intended to safeguard natural resources. Growing awareness of the interdependence of ecosystems and the imperative of sustainable management of natural assets fostered an emphasis on the long-term ecological consequences of human decisions. This critical alteration in outlook provided the conceptual underpinnings for contemporary federal environmental governance in the United States.

The Emergence of Environmental Awareness in the 20th Century

During the 20th Century, public perception of environmental problems underwent a profound and far-reaching alteration. This shift emerged from the simultaneous forces of rapid industrial expansion, advances in technology, and a dawning awareness of the intrinsic links between human behaviour and the biosphere.

The rapid growth of urban centres and factory production brought about stark evidence of air and water pollution, the swift exhaustion of natural resources, and the widespread degradation of habitats, thereby elevating ecological concerns to the forefront of national and international agendas. Influential voices—including Rachel Carson, whose pivotal 1962 volume *Silent Spring* documented the destructive effects of synthetic pesticides on living organisms—illustrated the delicacy of ecological balances and the potential repercussions of careless human intervention.

Environmental concern, moreover, emerged in dialogue with parallel social awakenings; campaigns for civil rights, feminist activism, and opposition to militarism wove together strands of thought about justice, empowerment, and the stewardship of the earth. This confluence produced a galvanising popular energy that was dramatically expressed in the inaugural Earth Day of 1970, when millions participated in rallies, plantings, and teach-ins designed to alert governments and fellow citizens to the need for conservation and sustainable practice. Scientific studies mapping atmospheric cycles, ecological thresholds, and the biosphere's carrying capacities, together with technological breakthroughs such as satellite remote sensing, provided empirical substantiation for the claims of activists and policymakers, thereby deepening the sense of urgency and calling for an integrated response to the rapidly unfolding environmental crisis.

The 20th Century saw a wide range of governmental responses to the growing environmental awareness, from initial scepticism to decisive regulatory actions. However, it also witnessed the emergence of critical institutions and legislation aimed at protecting natural systems, laying the groundwork for modern regulatory frameworks. Concurrently, the burgeoning environmental movement sparked a significant shift in societal values, with an increasing number of citizens incorporating sustainability into their daily lives. This cultural transition, marked by increased concern for conservation, the expansion of recycling programs, and the shift towards renewable energy sources, signalled a broader societal commitment to responsible environmental stewardship. Understanding these developments in the context of the 20th Century provides a more nuanced view of the origins of current environmental challenges and the ongoing quest for effective and sustainable solutions.

Landmark Legislation: The Clean Air Act and Beyond

The passage of the Clean Air Act in 1970 marked a significant milestone in the federal regulatory framework for environmental protection, demonstrating the nation's dedication to mitigating the adverse effects of airborne pollutants on public health and ecosystems. By empowering the newly established Environmental Protection Agency to set and enforce national ambient air quality standards, the Act established a comprehensive system for regulating emissions from various sources. Subsequent congressional reauthorisations and targeted amendments have reinforced the need for increasingly stringent and adaptable regulatory tools. These legislative changes have introduced strict performance standards for specific sectors, mandated the adoption of control technologies, refined compliance and enforcement procedures, and set innovative standards for emerging pollutants. The Clean Water Act has amplified the legacy of the Clean Air Act, the Endangered Species Act, and the National Environmental Policy Act, each contributing to a robust statutory framework for environmental protection. Together, these laws have facilitated a comprehensive regulatory system that integrates air, water, and habitat protection and promotes a precautionary approach to uncertain ecological risks. The coordinated operation of these statutes underscores the importance of addressing the interconnected and increasingly complex nature of modern environmental challenges.

Moreover, the body of statutes inspired by these landmark measures has encouraged cooperative engagements among federal, state, and local agencies, private-sector actors, and civic organisations, nurturing a durable ethic of stewardship and accountability toward the natural world. While the United States faces an evolving set of environmental challenges, the Clean Air Act and its legislative offspring furnish an enduring vindication of anticipatory regulation capable of slowing ecological decline and promoting the health and security of both current constituents and the generations that follow.

The Rise of the Environmental Protection Agency

Amid mounting environmental alarms and vigorous public petitions for remedy, the federal government, in 1970, founded the Environmental Protection Agency. This formative act constituted a decisive realignment toward a coordinated and unified scheme of ecological governance at the national level. Charged with the integration of dispersed programme and bureau missions relating to the preservation of public health and ecological integrity, the EPA assumed the central stewardship once scattered among an array of agencies, en-

CLIMATE CHANGE IN AMERICA 91

abling a more coherent and rigorous enforcement of environmental standards.

The establishment of the U.S. Environmental Protection Agency marked a decisive commitment of the federal government to manage a broad spectrum of environmental challenges through a single, cohesive entity. By integrating previously segmented functions—ranging from air and water quality surveillance to pollution abatement and applied environmental science—the Agency emerged as a formidable centre of regulatory oversight and policy execution. Its charge extends beyond limiting contaminant emissions and enforcing statutory provisions; it also encompasses the promotion of scientific investigation and the cultivation of collaborative relationships among federal, state, tribal, and local jurisdictions.

From the moment of its founding, the EPA has been a defining force in the formulation and institutionalisation of national environmental policy. By operationalising landmark statutory achievements such as the Clean Air Act, the Clean Water Act, and the Endangered Species Act, the Agency has enabled demonstrable progress in the safeguarding of air, water, and biodiversity. Its reach, however, is not confined by U.S. borders; the Agency actively contributes to multilateral efforts aimed at confronting transnational environmental challenges, thereby reinforcing the global dimension of its mandate.

The creation of the EPA initiated a transformative shift in public consciousness and institutional accountability concerning the nation's environmental affairs. By institutionalising mechanisms for citizen inquiry and comment, the Agency has enshrined transparency and participatory governance in environmental decision-making, thereby embedding democratic principles within the stewardship of natural resources. Over time, the EPA has come to be perceived as the institutional embodiment of civic engagement in the conservation of air, water, and ecosystems.

That said, institutional tensions and public debate have punctuated the Agency's trajectory. Regulatory initiatives, allocation of field and laboratory resources, and the perennial negotiation between economic competitiveness and ecological integrity have all been subjected to rigorous examination by stakeholders and the courts. In response, the EPA has recalibrated its methodologies, embraced adaptive policy techniques, and fortified its scientific underpinnings, thereby manifesting an institutional capacity for renewal under conditions of political and fiscal adversity.

A closer examination of the EPA's historical trajectory reveals that its foundational accomplishments have decisively shaped the architecture of environmental governance in the United States. The Agency's iterative rule-making, enforcement, and capacity-building initiatives have reconfigured the national policy contour and have created an enduring platform for confronting both the unpredictable threats and the strategic opportunities presented by a rapidly changing ecological and technological landscape in the 21st Century.

Key Amendments and Legislative Milestones

The sequence of Amendments and legislative breakthroughs in federal environmental policy has been instrumental in refining the United States' strategy for resource conservation and the mitigation of ecological harm. Over successive decades, key legislative interventions have fortified regulatory frameworks, fostered sustainable practices, and curtailed the deleterious consequences of human activity on the biosphere. The National Environmental Policy Act of 1970 stands as the first of these watershed initiatives; it created the Council on Environmental Quality and instituted the requirement that federal agencies conduct thorough environmental impact evaluations for all significant undertakings. The Clean Air Act Amendments of 1990 that followed introduced rigorous, technology-forcing commands and ambient standards that compelled a broad spectrum of sectors to lower pollutant releases. The 1973 enactment of the Endangered Species Act introduced a binding federal commitment to the conservation of imperilled fauna and flora and designated ecosystems, thereby embedding the goal of biodiversity preservation into the warp and woof of national law. The Resource Conservation and Recovery Act of 1976 provided a comprehensive strategy for the management of hazardous waste, coupling the regulatory imperative of safe disposal with the proactive goal of minimising waste generation and environmentally harmful practices.

These legislative benchmarks have decisively moulded federal environmental policy, while simultaneously informing international debates on sustainable growth and responsible environmental governance. The promulgation and subsequent enforcement of the Safe Drinking Water Act of 1974, in particular, reaffirmed the nation's dedication to the universal provision of clean and potable water. Together, the subsequent amendments and landmark statutes collectively illustrate the deepening national recognition of, and commitment to, environmental conservation and prudent resource stewardship. Confronted today with multifaceted ecological pressures, practitioners and policymakers must continually reference the historical context provided by these actions to assess present policy efficacy and to formulate resilient, forward-looking governance instruments.

Contemporary Challenges in Federal Environmental Policy

One of the key challenges in federal environmental policy is the delicate balance between economic growth and ecological stewardship. Industries, often well-resourced and organised, exert significant influence on the regulatory framework, which can sometimes compromise sustainability objectives. This necessitates elected officials to negotiate compromises that may dilute these objectives. Moreover, organised efforts to repeal, suspend, or substitute existing regulations introduce uncertainty and risk. To counter these challenges, governing institutions must strengthen existing standards and invest in regulatory structures that anticipate and mitigate the risk of retrogression, while also identifying areas where scientific evidence justifies stricter, evidence-based protections.

One of the most pressing challenges in federal environmental policy is the integration of responses to climate change within the governance structure. The increasing need for both mitigation and adaptation necessitates a proactive, anticipatory framework that acknowledges the interconnectedness of ecological, social, and economic domains. Operationalising climate resilience in federal policy requires an understanding of the evolving nature of environmental hazards, thereby highlighting the need for adaptive governance and flexible, evidence-driven decision-making. The growing civic engagement and demand for decisive climate policies further underscore the responsibility of legislators to pursue reforms that promote sustainability and reduce greenhouse gas emissions.

Another significant challenge in federal environmental policy is the imperative of environmental justice. Concentrated environmental harms disproportionately affect historically marginalised populations, underscoring the need to address structural inequities in access to clean air, potable water, and vital natural resources. Incorporating environmental justice into policy instruments requires a sustained commitment to rectifying past injustices and safeguarding the welfare of the most exposed and vulnerable. Achieving justice in environmental governance necessitates a comprehensive reassessment of existing regulatory regimes to ensure that protections are equitably distributed across diverse socio-economic and geographic contexts.

The contemporary pace of technological evolution, combined with the escalating emergence of environmental contaminants and the continuing trajectory of urban expansion, compounds the complexity of federal regulatory decision-making. Advances in detection, combined with the internationalisation of supply chains and biomes, now require a federal approach to accommodate the dynamic relationship between ecological disruption and human health. Consequently, contemporary federal policy must prioritise regulatory agility and constructive engagement with emerging science to remain responsive to both the trajectory of new environmental risks and the necessity of protecting human and ecological health.

Environmental Policy Under Different Administrations

Throughout American history, federal environmental policy has exhibited significant and often abrupt changes in response to the differing priorities and values of successive presidential administrations. Each administration has recalibrated the regulatory apparatus, the pace of enforcement, and the interpretation of statutory mandates, guided by the interplay of public opinion, the prevailing economic environment, and the federal government's emerging international obligations.

Initial statutory accomplishments, including the National Environmental Policy Act of 1969 and the formation of the Environmental Protection Agency under President Nixon, created the procedural and institutional foundation of modern environmental governance. Subsequent administrations have refined and diversified this foundation: President Carter elevated the prominence of energy conservation, and President Reagan elevated the role of market mechanisms, institutionalising regulatory flexibility and regulatory negotiation while delegating more decision-making to the states and private sector.

The Clinton administration signalled an era of reinvigorated attention to environmental governance, exemplified by the Climate Action Plan and the Clean Water Action Plan, which aimed at comprehensive approaches to climate and water resources, respectively. The succeeding George W. Bush administration framed its strategy as an effort to negotiate the twin imperatives of ecological integrity and economic vitality, introducing the Clear Skies Initiative to address stationary source emissions and the Healthy Forests Initiative to recast forest management with an emphasis on wildfire risk reduction and timber harvests.

Under President Obama, the federal apparatus once again elevated climate action, with the Clean Power Plan directing emissions reductions from the electricity sector and the United States endorsing the Paris Agreement as a multilateral cornerstone of global climate governance. The Trump administration then enacted a broad suite of deregulatory measures that included the rescission of the Clean Power Plan, the withdrawal from the Paris Agreement, and the recalibration of numerous permitting, emissions, and site-control standards.

The current Biden administration, characterised by multi-dimensional climate and environmental justice ambitions, has re-entered the Paris Agreement, announced a suite of clean-energy infrastructure investments, and employed executive authorities to weave climate considerations into the fabric of federal procurement, environmental review, and grant-disbursement processes.

This sequential unfolding of federal environmental policy highlights the variable constellations of statutory, market, and administrative instruments employed by successive administrations, revealing how policy evolution at the national level reflects broader socio-political contests and the recalibration of federal goals in the face of shifting scientific, economic, and electoral realities.

Judicial Influences on Environmental Legislation

Judicial influences significantly shape environmental legislation through the courts' interpretations of statutory and constitutional provisions. Within the statutory framework, judges effectively delimit the boundaries of regulatory jurisdiction, oversee the enforcement of statutory mandates, and resolve conflicts arising from environmental protection efforts. A particularly consequential judicial contribution is the refinement of legal doctrines that broaden the adoption and enforcement of environmental law. The standing doctrine, for example, has enabled both citizens and organisations to litigate environmental grievances, thereby reinforcing public participation and governmental accountability. Moreover, judicial rulings have delineated the constitutional and statutory authority of both federal and state governments to regulate activities that affect natural resources, as well as to safeguard air, water, and endangered species. The courts have further confronted intricate problems—including interstate pollution and cross-border ecological damage—thereby establishing precedents that inform legislative drafting, regulatory strategy, and intergovernmental coordination on environmental matters.

The judiciary plays a significant role in shaping environmental law, providing a robust legal framework. Court decisions have introduced equitable and fair principles into environmental governance, safeguarding the public interest and advancing environmental justice in litigation. Judicial analysis of statutory intent and scrutiny of regulatory actions have provided uniform applications of environmental statutes that remain faithful to their protective and sustainable objectives. Through careful examination and reasoned judgement, the courts have advanced the substantive doctrine of environmental law, creating a resilient framework capable of confronting ecological crises while encouraging prudent management of natural goods. Judicial interventions ensure that statutory regimes remain flexible, evolving in tandem with society's changing priorities and ecological realities, and sustaining a continuous exchange among law, scientific understanding, and the public interest.

Interagency Collaborations and Initiatives

Given that contemporary environmental challenges are frequently multiple levels and interrelated, coordinated interagency efforts are indispensable for aligning policy, action, and scientific knowledge within the broader framework of public governance. By merging statutory mandates, technical competencies, and budgetary authorities, federal agencies can fashion policy architectures that are not only coherent but also capable of addressing the full complexity of environmental conservation, spanning terrestrial and marine dimensions, pollutant control, habitat protection, and climate adaptation. Such collaborations encourage a systems-oriented perspective that transcends sectoral silos and recognises the interdependencies among land, air, and water resources.

A salient illustration of the efficacy of interagency collaboration is the initiative orchestrated by the National Oceanic and Atmospheric Administration, in conjunction with the Environmental Protection Agency and the U.S. Fish and Wildlife Service. This partnership confronts the multifaceted challenges that climate change poses to coastal ecosystems and human settlements. By conducting empirical research, permitting review, and hands-on ecosystem restoration, the agencies generate a decision-support framework that both shields sensitive coastal habitats and reduces the vulnerability of adjoining populations to phenomena such as accelerated sea-level rise and intensified storm vectors.

Additionally, the partnership between the U.S. Department of the Interior and the Department of Agriculture exemplifies how coordinated action across federal agencies can advance both sustainable land management and the conservation of biodiversity. By pooling specialised skills in forestry management, watershed stewardship, and habitat restoration, these departments reconcile the often competing imperatives of productive land use and the safeguarding of natural ecosystems on both federal holdings and broader rural terrains. Such integrative, cross-disciplinary efforts are critical for cultivating resilience and enduring sustainability under shifting climatic and environmental pressures.

The benefits of these interagency collaborations are further magnified by their international dimension, in which U.S. federal agencies align with foreign partners and multilateral institutions to confront environmental challenges that transcend national boundaries. The Department of State, in coordination with the United Nations Environment Programme, has thus catalysed cooperative measures on issues ranging from transboundary air pollution and the conservation of marine biodiversity to the management of hazardous substances. These transnational cooperative frameworks not only reinforce U.S. leadership in shaping a rules-based international environmental order but also deliver effective, coordinated responses to the increasingly interconnected suite of global ecological challenges.

Looking ahead, the effectiveness of interagency collaborations will hinge on the establishment of streamlined communication pathways, interoperable data platforms, and cohesive decision-making frameworks. Concurrently, cultivating a culture of mutual respect and formal recognition among agencies is essential to transcend bureaucratic impediments and synchronise priorities around shared environmental objectives. By leveraging the complementary capacities of the diverse federal agencies, the nation can develop holistic, forward-thinking, and sustainable strategies to preserve our natural heritage and to proactively address the dynamic challenges of the environmental future.

Reflections on Past Lessons for Future Policies

The cumulative history of federal environmental policy in the United States reveals recurring patterns that merit introspective examination as we chart future directions. The evolution of interagency coordination, coupled with statutory and administrative milestones, has forged a governance architecture that has both taught and achieved. Such examination must move beyond an inventory of laws and regulations, however, to interrogate the socio-cultural ethos, emergent technologies, and shifting global interdependencies that together have refracted national policy.

Above all, the record affirms the merit of anticipatory governance. Decades of environmental history reveal that reliance on ex-post corrective measures has repeatedly escalated public health burdens and fiscal liabilities. Strategic review of earlier episodes affords a reminder that investment in risk assessment, preventative regulation, and public preparedness can substantially curtail future liabilities. Equally necessary is the realisation that genuine environmental stewardship must operate on a timeline that outlasts electoral cycles and market fluctuations; durable policy architecture requires statutory and institutional scaffolding that resists the pull of short-term political expediency.

A second vital insight is the inescapable reality of ecological interdependence. Past efforts—whether targeting air quality, water resources, or hazardous waste—quickly disclosed that measures confined to isolated media or sectors frequently precipitate unforeseen consequences elsewhere in the web of life. The record therefore counsels a systems-level sensibility, one that institutionalises cross-media and cross-jurisdictional analytical practices as core components of policy design. When coupled with adaptive management practices that incorporate new data and evolving scientific understanding, such an approach can increase policy resilience in the face of complex and often cumulative environmental change.

The gradual acceptance of holistic and integrated environmental management has emerged from accumulated experience, emphasising that enduring solutions depend on relationships that bridge sectors and disciplines. This recognition confirms that environmental challenges do not occur in isolation but present themselves as complex, interlinked systems. Looking back from this vantage point, we are encouraged to recognise the intricate character of the biosphere and to commit ourselves to systems of cooperation and interdisciplinarity that the present moment demands.

Change in environmental policy has always moved in concert with public sentiment, social movements, and technological progress. By examining how policy has been informed and shaped by these social currents, we gain a clearer picture of how emerging norms might steer the next generation of statutes and regulations. Acknowledging that public preferences and collective values are not mere context but essential inputs to policy design prompts each governance cycle to remain flexible and reflective. Simultaneously, the speed of technological innovation obliges policymakers to adopt a proactive posture, integrating the present and anticipated capabilities of novel tools into the regulatory baseline.

The review of the federal policy record, therefore, amounts to a composite of lessons that, when articulated and integrated, can provide a sturdy frame for the continuing evolution of sound, effective, and just environmental governance.

The values of anticipatory insight and relational awareness, together with the lively mutual shaping of policies and social practices, provide the orienting principles for the continuing evolution of environmental governance. Consequently, the historical analysis reaches an outcome that transcends reflection; it articulates well-defined trajectories leading toward an enduring and adaptable ecological future.

7
State and Local Leadership in Climate Action

State and Local Leadership

Rising climate impacts have pushed state and local authorities to the forefront of the global sustainability agenda. No longer seen as mere conduits for federal guidance, these governments have become the laboratories of climate policy, demonstrating that effective change can start at the community level. Historically, the need for local engagement became stark after national climate programmes encountered legislative gridlock and the need for tailored, context-sensitive solutions became clear. This section examines the current state of local and state climate leadership, illuminating the diverse incentives—economic, social, and political—that drive action, the key milestones that have built momentum, and the expanding policy frameworks, from ambitious renewable energy standards to innovative zoning reforms, that equip them to chart the nation's path toward a low-carbon, climate-resilient future.

Historical Background and Legislative Framework

The emergence of climate leadership at the state and local levels has arisen out of shifting environmental awareness, evolving policy trajectories, and persistent public advocacy. Initially, public concern was largely confined to acute, place-specific pollutants and the depletion of localised resources. Attention to climate change itself crystallised only in the decades following World War II, at which point legislators began to translate science into law at the federal, state, and municipal levels.

The federal architecture was decisively strengthened by the Clean Air Act of 1970 and the 1970 establishment of the Environmental Protection Agency. Those instruments created a supervisory framework empowering state and municipal governments to craft regulations consistent with national objectives. Regions, in turn, confronted the looming spectre of climate change through an eclectic mix of legal instruments, adapting federal parameters to their geographic, economic, and social particularities.

Today's legislative landscape relies on a heterogeneous assortment of statutes, administrative rules, and executive proclamations. Many states have internalised high targets for greenhouse gas reduction, enforced renewable portfolio mandates, and devised programmes to incentivise energy efficiency. Moreover, a growing number of jurisdictions have adopted cap-and-trade systems, alongside carbon markets, intended to limit emissions while simultaneously promoting a resilient, low-carbon economy.

On the local governance front, municipalities and counties have enacted innovative ordinances governing building codes, land-use planning, and transportation networks, thereby affirming the critical contribution of cities to lowering greenhouse-gas emissions and improving adaptive capacity to climate extremes. In tandem, several local governments have formed multi-jurisdictional coalitions that facilitate the exchange of successful strategies and magnify the overall effect of their climate-policy endeavours.

A historical survey of climate governance at state and local scales reveals a dynamic interplay of collaboration, trial-and-error learning, and responsive policy modification. Legislative instruments have gradually expanded to encompass an increasingly inclusive array of environmental domains, embedding climate-related criteria in agriculture, spatial planning, and public-health regulations. Fuelled by a shared sense of urgency and constructive intergovernmental collaboration, these subnational authorities are charting an ever-evolving course for climate policy, thereby advancing the nation toward a durable and sustainable future.

Case Studies: Innovative Policies and Programs

To understand effective climate action in practice, we must analyse cutting-edge policies and programmes adopted by states and municipalities. This section highlights several illustrative case studies featuring novel methods for both mitigation and adaptation. A leading example is California's cap-and-trade system, which assigns a monetary cost to carbon emissions, thereby motivating industries to cut greenhouse gases while creating a revenue stream for complementary environmental initiatives. In a comparable vein, New York's Clean Energy Standard obligates a phased transition toward renewable power, demonstrating a vigorous state-level commitment to climate mitigation. At the municipal tier, Portland, Oregon, distinguishes itself through a multifaceted sustainable transportation agenda, which encompasses a dense network of bike lanes, the expansion of EV charging stations, and financial incentives for electric vehicles, all of which contribute to lower emissions and healthier urban air. Added to these successes, Norfolk, Virginia, exemplifies effective community-based resilience to flooding: by fostering alliances among local government, nonprofit agencies, and neighbourhood residents, the city has implemented adaptive measures that enhance both infrastructure and social solidarity in a context of rising seas.

We will additionally examine a range of smaller municipalities throughout the Midwest that have integrated energy efficiency and renewable energy initiatives, frequently in partnership with regional businesses and nearby colleges or universities. These case studies will reveal the varied, locally adapted methods that states and communities have employed to address climate change, demonstrating the likelihood that effective practices can be reproduced and expanded. By scrutinising the governing structures, hurdles, and measurable results of each project, our analysis seeks to portray the evolving territory of climate policy execution beyond the federal level and to emphasise the essential, proactive contribution that subnational governments make in advancing sustainability and enhancing community resilience.

The Role of State Governments in Climate Advocacy

State governments constitute the first meaningful tier of climate advocacy within the American polity. Functioning as laboratories of democracy, they are empowered to enact emissions mandates, foster renewable energy markets, and cultivate norms of environmental stewardship. Many states have self-imposed, science-informed carbon-neutrality deadlines, financed renewable installations, and activated coalitions of municipalities, businesses, and civil organisations. Positioned between shifting federal levers and community-based outcomes, they translate broad national goals into measurable zoning codes, procurement strategies, and incentive auctions. Legislative caucuses, gubernatorial directives, and interstate compacts collectively amplify state voices in national policy conversations. Regulatory agencies, environmental protection departments, and offices of energy constantly calibrate permitting protocols, supply modelling, and outreach to underserved populations.

These agencies author the technical memoranda that guide emissions inventories, and they broker memoranda of understanding that align utility decarbonisation plans with equity mandates. State governments underwrite core research in advanced battery chemistry, energy-efficient buildings, and flood-resilient infrastructure, funding pilot arrays, granting seed capital, and mandating climate considerations in capital budgeting. In short, state governmental apparatuses represent the most immediate and adaptable vehicle for calibrating climate policy to the geography, economy, and political milieu they share with the governed.

Through deliberate promotion of innovation and entrepreneurial ecosystems, states can expedite the global shift to a low-carbon economy. Furthermore, advocacy led by state governments transcends the drafting and enforcement of domestic statutes. States cultivate diplomatic networks, engaging both fellow American jurisdictions and international partners to share effective practices, align regulatory frameworks, and jointly confront transboundary environmental dilemmas. Such cooperative endeavours illuminate the intrinsic connectivity of climate challenges and reinforce the necessity of multilateral governance anchored at the state tier. The climate advocacy function of state governments is layered and requires anticipatory vision, cooperative networking, and sustained, long-range dedication. As the intricacies of climate governance evolve, it is crucial to recognise and amplify the ability of states to act as decisive engines of environmental advancement.

Local Government Initiatives: Cities Leading the Charge

At the cutting edge of climate action, cities are rapidly becoming the laboratories of experimentation that the planet urgently needs. Their stature in the journey toward sustainable development and climate resilience is unmatched. Grounded in the daily realities faced by citizens, local governments are turning the mounting evidence of climate upheaval into decisive policies and actionable programmes designed to quell environmental hazards and embrace the imperatives of a shifting climate. Nationwide, municipalities are converting regulatory authority and spatial advantages into harvestable opportunities, championing low-carbon economies, scaling solar and wind deployments, and committing to ever-tightening emissions caps.

Collaborating with private enterprises, universities, and resident networks, cities are translating local strengths into global relevance, altering the narrative of who can lead climate governance. Their success is a living template, encouraging provinces, states, and fellow nations to adopt similar vigour. Moreover, local initiatives are finely calibrated to the geographical and socio-economic contexts they serve, acknowledging that uniform solutions overshoot their targets. Whether by embedding mixed-use development, retrofitting municipal and private buildings for energy performance, closing material loops through waste minimisation, or embedding climate in urban design, cities are executing a comprehensive and multi-sectoral arsenal of strategies that shrink carbon footprints and crystallise climate-resilient futures.

Integrating climate-adaptation strategies into municipal planning is empowering cities to take the lead in managing climate risks proactively. More than passing regulations, city-led outreach campaigns are engaging citizens, raising environmental awareness, and encouraging sustainable living that supports local climate projects. Through public-private alliances and collaboration across sectors, municipalities are leveraging every opportunity to enact lasting change while strengthening critical infrastructure against climate impacts. As cities become indispensable players in climate action, sharing best practices and documented lessons among them is essential. Such exchanges foster innovation and enable successful strategies to be replicated, speeding the global shift to a low-carbon, sustainable future. With local governments actively involved, the foundation is being built for urban environments that are resilient, adaptable, and environmentally conscious, providing a compelling model and a source of optimism in the global climate battle.

Interstate Collaborations and Regional Compacts

Interstate collaboration and regional compacts are becoming critical instruments for mitigating climate change across the United States. Realising that climatic and environmental phenomena regularly transcend state lines, multiple jurisdictions are now negotiating shared programmes that synchronise their regulatory, technical, and financial resources. Such strategic partnerships are immediately necessary for enhancing regional resilience, encouraging low-carbon economic sectors, and protecting interlinked ecosystems and hydrological networks.

A prominent illustration of this approach is the Regional Greenhouse Gas Initiative (RGGI). Consisting of multiple Northeast and Mid-Atlantic states, RGGI establishes a legally binding cap-and-trade system for carbon dioxide pollution from power plants. The auction of emissions allowances is returned, in part, to fund efficiency and renewable programmes. RGGI has achieved both declining emissions and economic reinvestments, clearly evidencing how collaborative governance across jurisdictions can fulfil emissions-reduction commitments while preserving regional integrated markets.

At the same time, the Western Climate Initiative (WCI) provides a broader platform for sub-national jurisdictions to adopt compatible cap-and-trade programmes, phase down short-lived climate pollutants, and expand clean-energy deployment. Composed of states in the United States and provinces in Canada, the WCI is linked to a standard set of monitoring, reporting, and verification procedures, allowing seamless allowance banking and trading across its evolving policy instruments. The WCI demonstrates that coordinated climate action can confer economic efficiencies, facilitate technology-sharing, and overcome the market fragmentation that otherwise weakens individual state efforts.

Interstate collaborations reach beyond the management of carbon emissions and energy policy. The Great Lakes-St. The Lawrence River Basin Sustainable Water Resources Agreement illustrates how states and provinces surrounding the basin can cooperatively govern one of the world's most vital freshwater resources. The agreement establishes a common vision and a series of shared stewardship practices that enable the jurisdictions to tackle the inseparable issues of water quality, quantity, and equitable availability while safeguarding the basin's ecological health.

Beyond the singular topic of water, these inter-jurisdictional partnerships provide a durable forum for the exchange of information, the dissemination of best practices, and the development of administrative capacity. Through joint scientific studies, coordinated data collection, and public-facing outreach, the states refine their understanding of regional climate patterns and anticipate future vulnerabilities. This shared knowledge allows each jurisdiction to tailor its adaptation policies and regulatory frameworks to the specific conditions facing its communities.

The promise of these collaborations, however, can be undermined by regulatory divergence, shifting political priorities, and unequal allocation of financial and human resources. Addressing these barriers calls for persistent dialogue, pragmatic compromise, and a culture of mutual trust among the partner jurisdictions. Transparent governance frameworks and robust, cooperative dispute resolution mechanisms remain indispensable for upholding the integrity and effectiveness of the initiatives.

In the years ahead, interstate collaborations and regional compacts will remain critical instruments for advancing meaningful climate action. By uniting a broad array of stakeholders and synchronising efforts around shared goals, these alliances can magnify the effects of discrete state and municipal measures, thereby paving the way to a more resilient, low-carbon future.

Challenges and Barriers to Effective Local Action

Local initiatives to address climate change confront an array of intertwined challenges that can stymie their momentum and thoroughness. A primary hurdle is the scarcity of financial resources. Many municipalities find it difficult to earmark the necessary budget allowances for programmes that mitigate greenhouse gas emissions or enhance climate resilience. Compounded by competing budgetary demands, these financial pressures frequently result in environmental concerns being sidelined.

Coordination challenges add further complexity. Mobilising a diverse set of municipal departments, public agencies, and civic organisations around a single, coherent climate action strategy is often fraught with difficulty. Differing institutional priorities, overlapping jurisdictions, and sometimes conflicting mandates can fragment efforts, complicating the pursuit of common, science-based objectives.

Political opposition, coupled with wavering public enthusiasm, can considerably complicate the adoption of robust climate policies at the municipal level. When prominent community figures publicly question or resist climate science, mobilising broad-based support for proposed measures becomes particularly difficult.

A further significant obstacle is the lack of uniform benchmarks and protocols for tracking and reporting on the progress of local climate initiatives. In the absence of well-defined indicators and standardised evaluation methods, jurisdictions struggle to measure the effectiveness of their actions or to make meaningful comparisons with peer communities.

Limited technical capacity among local elected officials and administrative staff also hampers credible climate action. Mastering the complex interplay of sustainability planning, emissions-reduction pathways, and resilience-oriented infrastructure requires specialised training that local governments may not always possess.

Finally, the volatility of funding sources and evolving policy landscapes creates a pervasive sense of uncertainty that can stymie sustained effort. When communities cannot rely on predictable, long-term financial and policy support, they often scale back or delay ambitious climate initiatives, fearing that future administrations may not reward sustained investments.

The global interdependence of climate systems ensures that even the most rigorous local initiatives remain susceptible to dynamics that originate far beyond their jurisdictions. This interconnection highlights the necessity of coordinated international, national, and regional partnerships, which ensure that locally tailored interventions are reinforced rather than undermined by wider trends and policies.

Nonetheless, local communities continue to demonstrate remarkable resilience and ingenuity. By candidly acknowledging the constraints imposed by external and systemic factors, local leaders are better positioned to refine their climate action strategies. Such candid assessments, coupled with targeted capacity-building, can elevate the effectiveness of initiatives and hasten the transition toward sustainable and climate-resilient futures across diverse geographies.

Engaging Communities: Bringing Climate Solutions Home

Involving communities actively in the fight against climate

change is indispensable for the creation of enduring and equitable solutions. As climate impacts manifest more intensely at the neighbourhood and regional levels, the collective ownership of adaptation and mitigation measures not only ensures that interventions reflect local realities but also builds investment in their success. This section delineates a spectrum of participatory methodologies, capacity-building practices, and co-design processes that facilitate meaningful community contribution to climate action planning and execution.

Community engagement starts with sustained education and outreach. Accessible, precise information about climate change, its expected impacts at the local scale, and the critical role of citizen participation in mitigation and adaptation is the foundation. Efforts should focus on conveying the feasibility of sustainable energy alternatives, energy and water conservation practices, and resilience strategies that residents can integrate into everyday life.

Equally important is the principle that community members must be integrally involved in the policymaking continuum. Participation in the design, deliberation, and evaluation of climate policies ensures that programmes can respond sensitively to the varied priorities and capacities within the population. When individuals believe that their voices shape decisions, they gain a sense of stewardship that enhances the durability and effectiveness of the resulting measures.

Beyond education and participatory policymaking, the cultivation of partnerships across a community's sectors is critical. Collaboration among local businesses, non-governmental organisations, academic institutions, and neighbourhood organisations can accelerate the identification and replication of pragmatic solutions. Such a network broadens the pool of information, financial resources, and technical expertise, enabling a coordinated, context-specific response to the climate vulnerabilities that each community uniquely experiences.

Effective communication and outreach must be designed to be inclusive and culturally responsive so that every segment of the community can fully participate and be empowered. By intentionally recognising and integrating diverse viewpoints, traditional ecological knowledge, and indigenous stewardship practices, climate strategies can be strengthened and decision-making can reflect the full array of community values and ways of life. Such an approach not only respects the community's heritage but also enriches the planning process by tapping into established, context-specific evidence of sustainable practices.

Successful engagement will intentionally move beyond isolated programmes, working to cultivate enduring capacity and long-term resilience among local populations. By fostering an enduring ethic of ecological stewardship and sustainable development, communities can generate momentum that carries climate solutions into the future. Ongoing, inclusive dialogue, cooperative action, and a shared sense of responsibility can collectively transform communities into proactive, empowered partners advancing the global agenda to limit warming and to adapt to the changes that are now unavoidable.

Measuring Success: Metrics and Evaluation

Turning to the pivotal question of evaluating climate action effectiveness, it is clear that an integrative framework for measurement and appraisal is indispensable. This part of the analysis will examine the complexities of constructing metrics that are both meaningful and context-sensitive, and will underscore the necessity of rigorous evaluations to determine whether policies and programmes are producing durable, positive climate outcomes.

Assessing the efficacy of mitigation and adaptation measures in a warming world demands an integrative and layered framework. On one level, mechanistic indicators—such as the trajectory of net greenhouse gas emissions, the fraction of final energy supplied by renewables, and the measured improvements in ambient air and surface-water quality—provide robust, quantifiable evidence of change. On an equally critical level, qualitative dimensions—such as the depth of public understanding, the breadth of stakeholder participation, and the fairness of outcomes across income and racial groups—must inform any judgement about the depth and longevity of climate progress.

The principal difficulty, then, is to construct context-sensitive evaluative protocols that adequately capture the full spectrum of consequences arising from climate interventions. To illustrate, an appraisal of a municipality's solar-plus-storage initiative must pursue, in concert, the observed savings in emissions, the distribution of energy cost savings across vulnerable households, and the programme's resilience to future climatic perturbations. An evaluative culture that acknowledges the fusion of environmental, social, and economic strands is thus essential to rendering a true and enduring account of success in climate governance.

Data collection and analysis represent foundational activities in this undertaking. The deployment of cutting-edge monitoring instruments, coupled with geographic information systems and integrated databases, allows for meticulous observation of vital performance indicators. Systematic examination of these data equips policymakers with evidence of the relative efficacy of alternative interventions, directs attention to underperforming sectors, and supports judicious planning of resource investments for upcoming climate programmes.

Assessment of climate action further requires active engagement and cooperation among a wide range of stakeholders. When government entities, civil society, industry representatives, and local communities participate in the evaluation, the resulting analysis is both broader and deeper. Such inclusive dialogue enhances transparency, reinforces accountability, and guarantees that the experiences and concerns of all interested constituencies inform the appraisal of policy and programme effectiveness.

Future Prospects and Emerging Opportunities

To summarise, measuring success in climate action requires a pluralistic methodology that merges quantitative and qualitative indicators, advanced data-intensive analysis, and participatory stakeholder processes. The development of sound appraisal metrics and the execution of rigorous evaluations are, therefore, critical to advancing climate interventions that are both effective and enduring, thus safeguarding ecological, social, and economic well-being for current and forthcoming generations.

Rich prospects for innovation, cooperative engagement, and substantial change characterise the outlook for climate action at the state and local levels. Continuous assessment and adaptation of our strategies for confronting the climate crisis compel us to anticipate strategic opportunities that can yield measurable progress. Within the sphere of subnational governance, several observable trajectories suggest that tomorrow's climate action will be forged here. At the forefront of technology and governance, progress in renewable energy sources, intelligent infrastructure, and sophisticated data analytics is enabling communities to enhance resilience while concurrently abating greenhouse gas emissions. Whether through the deployment of microgrid systems or the introduction of blockchain-based carbon accounting mechanisms, states and municipalities can now mobilise frontier technologies to advance the sustainability agenda in practical, quantifiable ways. Equally, the growing mandate to embed environmental justice into the core of climate initiatives is gaining traction. Policymakers are increasingly attuned to the fact that vulnerable populations and frontline communities bear the heaviest climate burdens, and their insights are being integrated into the formulation and evaluation of state and local climate strategies.

   Prioritising fairness in climate initiatives deepens their moral legitimacy while reinforcing the adaptive capacities of entire communities. A promising avenue for further advancement is the expanding landscape of public-private partnerships. When governments, corporations, and local organisations pool expertise, resources, and networks, they can propel the economy away from fossil fuels far more rapidly. Through strategic reform of public purchasing, targeted financial incentives, and collaborative research, these alignments can deliver measurable gains for sustainability. In parallel, the sustained engagement of youth and community-led movements continues to serve as a powerful catalyst for localised action. The commitment, creativity, and clarity of purpose exhibited by younger advocates have already reshaped climate discourses, generating net-zero pledges and broadening public concern. Capturing this energy while ensuring that diverse perspectives inform policy design can refresh the climate strategies of municipalities and states alike. Moving forward, it remains essential to acknowledge the intertwined character of global challenges and openings. By promoting interdisciplinary partnerships and reciprocal sharing of best practices, subnational authorities can assert their agency within the broader climate effort. A comprehensive orientation that integrates social equity, economic viability, and ecological integrity is indispensable in generating a future that is both resilient and sustainable.

In the end, the landscape of state and local climate initiatives is rich with avenues for constructive, far-reaching amendments. If anchored by deliberate planning, unyielding resolve, and an ethos of joint enterprise, these new openings can become the engines for substantive progress, ensuring a resilient, just, and ecologically sound world for every individual.

8
US Engagement in Global Climate Action

A Comparative Analysis of Republican and Democratic Administrations

International Cooperation and Diplomacy

The United States' role in global climate governance is closely intertwined with its engagement in international cooperation and diplomatic initiatives. Across successive administrations, Washington has adopted a range of strategies to address the intricacies of climate diplomacy and to foster multilateral collaboration. Within the global climate arena, the United States has participated actively in the United Nations Framework Convention on Climate Change (UNFCCC) and its periodic conferences, particularly contributing to the negotiating process that culminated in the 2015 Paris Agreement. These diplomatic efforts have permitted the United States to help shape international climate architectures, champion legally binding and non-binding emissions reduction pledges, and advocate for the creation and capitalisation of financial instruments to assist vulnerable nations.

Republican administrations have consistently approached international climate cooperation with greater circumspection, foregrounding economic vitality and expressing doubts about the long-term efficacy of global accords. This predisposition reached a high-water mark in 2017 with the decision to withdraw from the Paris Agreement, a move that precipitated international unease regarding the credibility of US climate leadership. The withdrawal was accompanied by pronounced reservations about climate finance fellowships, demands for equitable burden-sharing, and a diplomatic posture that emphasised reciprocal gains rather than collective responsibility.

Democratic administrations have consistently pursued the re-engagement of international partners, the deepening of climate cooperation, and the channelling of diplomacy into the reinforcement of global emissions-reduction and resilience commitments. The Obama administration's orchestration of the Paris Agreement, coupled with bilateral accords with critical nations, showcased a deliberate effort to restore US primacy in climate diplomacy. The Biden administration, by rejoining the Paris Agreement and hosting a Leaders Summit on Climate in 2021, reaffirmed the administration's sustained focus on multilateralism and collaborative action in the face of the climate emergency.

US engagement in international climate cooperation has, in turn, been moulded by domestic political currents, the prevailing zeitgeist of public opinion, and the progressive refinement of global environmental governance. The negotiation of diplomatic initiatives has reflected the negotiation of domestic imperatives—economic growth, energy security, and environmental stewardship—while influencing the capacity to lead on and shape multilateral climate discourse. Perceptions of US leadership in diplomatic fora, and, by extension, the successful pursuit of a global climate agenda, have linked the country's soft-power assets to its capacity to forge and sustain enduring political coalitions with a diverse array of states.

When assessing how the United States contributes to international cooperation and diplomacy for global climate action, it is crucial to situate this engagement within the evolving interplay of geopolitical interests, economic interdependencies, and collective environmental threats. The US's ability to galvanise international momentum for climate action and to cultivate cooperative frameworks will remain a decisive element for confronting the multifaceted, transnational character of the climate emergency.

Amid the escalating imperative to curb global warming, the United States continues to occupy a central diplomatic position. Its capacity to steer multilateral climate negotiations, to frame global governance instruments, and to partner across a diverse constellation of states can accelerate collective mitigation and adaptation efforts. Such engagement must encompass the major economies that account for the bulk of emissions, emerging markets that are charting low-carbon development, and the most climate-vulnerable nations that bear the disproportionate losses. Sustaining diplomatic credibility will therefore hinge on trust-building, on facilitating the transfer of clean technologies, and on mobilising predictable, climate-sensitive financial flows, all of which are indispensable to confronting the climate challenge that transcends national borders.

The effectiveness of American leadership in international climate diplomacy hinges, crucially, on its resolve to translate aspirational rhetoric into robust domestic climate policy. When domestic policy frameworks mirror the pledges made on global platforms, the United States amplifies its credibility and capacity to steer the multilateral agenda. Concrete measures include accelerating the transition to a low-carbon energy sector, prioritising the resilience of infrastructure, and embedding climate considerations into trade and foreign policy frameworks. Simultaneously, American diplomacy must extend beyond conventional state-to-state interactions by enlisting governors, business leaders, and civil society actors, whose innovative practices and commitments can independently drive global cooperation on climate mitigation and adaptation.

In the sphere of international climate finance, the US must marshal both public and innovative private-sector capital to underwrite transition and resilience efforts in developing countries. By funding resilience-building infrastructure, enabling technology diffusion, and supporting workforce training programmes, the
United States not only alleviates global vulnerability but also solidifies its negotiating leverage. Participation in the Green Climate Fund, alongside robust bilateral and multilateral development finance, will underscore American reliability and reinforce a multilateral norm of climate solidarity. Together, these actions forge a durable foundation on which the US can advocate for ever-more ambitious global emissions reduction trajectories.

In its current push to reaffirm global climate leadership, the United States must prioritise collaboration with other large emitters and rapidly developing economies. This effort should centre on deliberate, high-level conversations with China, the European Union, India, and other pivotal stakeholders to harmonise climate objectives, strengthen transparency, and encourage the deployment of forward-looking technologies. Simultaneously, the United States can exert critical influence by championing climate initiatives in established multilateral settings, including the G7, G20, and relevant regional groupings, generating synchronised action and elevating worldwide climate benchmarks.

The effects of American climate diplomacy extend well beyond conventional strategic rivalry. They are shaped by the urgent, universal necessity of limiting climate disruption and ensuring a viable planet for present and future generations. When diplomacy is deployed to create consensus, sustain momentum, and induce coordinated responses, the United States can make a meaningful contribution to the global climate enterprise. This approach requires harnessing the full spectrum of American strengths—political weight, scientific knowledge, advanced technologies, and financial resources—to mobilise international collaboration and accelerate the global shift toward a resilient, low-carbon future.

The United States' engagement in global climate diplomacy and multilateral negotiations signals an enduring commitment to collective responsibility in confronting the climate emergency. Through its capacity to cultivate consensus, reconcile disparate viewpoints, and advocate for transformative technologies, the United States remains pivotal to the intricate architecture of global climate governance. Concurrently, the nation's domestic drive to decarbonise its economy, expand a resilient green workforce, and uphold robust environmental stewardship must be harmonised with international collaboration. Together, these imperatives will define the United States' capacity to influence an equitable, sustainable, and prosperous trajectory for the planet and its future generations.

9
Balancing Mitigation and Adaptation Strategies

Mitigation and Adaptation

Mitigation and adaptation together constitute the dual framework within which global society is responding to climate change. Mitigation is concerned with the reduction or outright prevention of greenhouse gas emissions and the human drivers of climatic disruption. Its domain of action is broad, including the wholesale shift to renewable energy, the optimisation of energy use, the promotion of sustainable land-management practices, and the development of carbon capture and storage technologies. The intention behind these measures is to halt the upward trajectory of greenhouse gases, thereby addressing the process that is fundamentally driving the alteration of the planet's climate.

Meanwhile, adaptation, a deliberate modification of natural and human systems, is a crucial aspect of our response to observed or anticipated climate change. This approach emphasises the advanced preparation necessary to reduce potential harm from shifting environmental conditions while simultaneously leveraging any advantageous transitions. Adaptation measures span a broad spectrum, including the design of climate-resilient infrastructure, the refinement of agricultural techniques capable of withstanding extreme events, and the establishment of proactive early warning systems against natural hazards. The central objective of adaptation is to lessen the exposure and susceptibility of both communities and ecosystems, thereby enhancing overall resilience and supporting sustainable development.

The primary difference between mitigation and adaptation resides in their points of intervention: mitigation targets the underlying drivers of climate change, whereas adaptation confronts the resultant impacts. Both strategies are critical to any holistic climate governance framework, yet their efficacy is maximised when they are synthesised into a unified policy apparatus. Effective mitigation efforts constrain the scale of potential climate impacts, thereby diminishing the residual risk that adaptation measures must address. Conversely, timely and well-resourced adaptation initiatives safeguard human and ecological systems from changes that are already locked in, thereby fostering an anticipatory culture of preparedness.

A holistic understanding of both mitigation and adaptation is essential for effectively addressing climate change, as it allows for the design of policies that capitalise on the complementary effects of each approach. Policymakers, sector actors, and communities therefore must engage in joint, iterative processes to create and enforce integrated climate policies that equitably distribute effort between mitigation and adaptation initiatives. When these initiatives are aligned with wider sustainable development objectives, societies are better positioned to realise climate-resilient trajectories that reduce greenhouse gas concentration, protect ecosystems, and simultaneously advance economic, social, and environmental justice.

Defining Mitigation Strategies

Mitigation strategies are core elements of any climate action framework that seeks to curtail the magnitude and effects of climate disruption. These strategies confront the drivers of change by reducing anthropogenic greenhouse gas emissions and by augmenting processes that can remove carbon dioxide from the atmosphere. One prominent mitigation tactic is the deployment of renewable energy technologies—solar, wind, and run-of-river hydro—replacing the combustion of fossil fuels. Complementing this shift, efficiency improvements in electricity generation, industrial processes, and residential energy use are indispensable for limiting the energy-related emissions that underpin climate change.

Land-use planning and forest stewardship occupy central roles in greenhouse gas mitigation. Strategic afforestation, reforestation, and responsible forestry increase carbon uptake and storage, effectively diminishing atmospheric CO_2 concentrations. Complementary efforts in sustainable agriculture and soil carbon management further curtail emissions by enhancing soil organic matter and minimising losses. In parallel, carbon capture and storage (CCS) technologies deliver critical options for heavy industries facing deep decarbonisation hurdles, enabling the secure confinement of CO_2 beneath the subsurface.

Successful mitigation, however, requires more than engineering and land management; it depends upon policy architectures that align incentives with emissions reduction. Carbon pricing regimes, whether in the form of explicit taxes or market-based cap-and-trade systems, send potent economic signals that prompt firms and households alike to adopt lower-carbon choices. Complementary regulatory instruments, including performance standards and technology mandates, reinforce these signals by accelerating the diffusion of clean energy and production practices, thus steering economies toward robust low-carbon futures.

Strategic planning of urban environments and associated infrastructures prioritises mitigation measures that decrease dependence on carbon-heavy transport modes and bolster building resilience to extreme climatic events. The conscious placement of urban green corridors alongside integrated sustainable water practices reduces heat-island phenomena and enhances carbon capture. Moreover, transnational collaboration, buttressed by institutional partnerships, remains vital to synchronise emission-reduction endeavours and amplify the effect of collective commitments.

Recognising and executing comprehensive mitigation portfolios remains essential for attenuating the chronic consequences of climate change and for protecting the welfare of both current and future populations. Through systematic articulation and systematic administration of these portfolios, communities may cultivate durable and adaptive trajectories that respond constructively to the planet's deepening ecological crises.

Exploring Adaptation Techniques

Adaptation techniques are pivotal elements of any comprehensive climate change response framework, prioritising resilience and preparedness amid ongoing and predicted environmental transformations. This section offers an in-depth examination of a suite of adaptation techniques available for limiting climate impacts on human and ecological systems.

1. Ecological Restoration: Restoring and safeguarding natural ecosystems—wetlands, forests, and grasslands—profoundly strengthens biodiversity and preserves crucial bio-geophysical processes. Through systematically rehabilitating degraded habitats, practitioners and policymakers bolster the capacity of ecosystems to absorb shocks, support flora and fauna, and buffer against extreme hydro-meteorological events.

2. Infrastructure Resilience: Embedding resilience-oriented design criteria in public and private infrastructure projects is essential for coping with dynamic climate conditions. These criteria may involve elevating critical facilities, retrofitting utilities and transportation corridors to withstand flooding and heat, integrating smart drainage and storage systems, and adopting green infrastructure elements, such as permeable pavements and urban forests, to mitigate urban heat islands and enhance microclimate regulation.

3. Agricultural Innovation: Resilience in agriculture is

CLIMATE CHANGE IN AMERICA

non-negotiable for global food security and the viability of rural economies. Strategies such as integrating climate-resilient crop varieties, dynamising rotation and intercropping systems, optimising irrigation with sensors and weather data, and adopting agroforestry systems that combine tree cover with crops and livestock improve adaptive capacity while concurrently enhancing soil carbon storage and conserving soil and water resources.

4. Community-Centric Strategies: Involving local populations in adaptation initiatives is not just a strategy, but a necessity. This approach leads to greater equity and empowers residents to strengthen their livelihoods against climate impacts. Community-oriented programmes might include awareness campaigns, participatory planning frameworks, and the adoption of measures finely adjusted to the distinct vulnerabilities, cultures, and resources of each locality.

5. The active involvement of communities is a key element in the success of adaptation strategies. Emerging Technologies: The application of new tools—early warning networks, remote-sensing platforms, and sophisticated climate models—can hold immense potential to enhance the ability of systems and societies to adapt. These technologies deliver rapid assessments of evolving hazards and support strategic choices that limit exposure and mitigate damage.

6. The continuous development and application of these technologies provide hope for a more resilient future. Flexible Governance: Constructing governance arrangements that can adjust across scales—from village councils to ministries—is essential to sustaining progressive adaptation.

Policies that prioritise flexibility, stakeholder inclusion, and rapid response embed resilience concerns in normative frameworks and secure their continuation in successive planning and regulatory cycles.

Together, these lines of inquiry highlight a broad and interconnected set of proactive options for confronting the layered pressures of climate change. When fused into cohesive adaptation frameworks, such options enable communities and nations alike to advance their capacity to prosper amidst a turbulent climate future.

The Interplay Between Mitigation and Adaptation

The interplay between mitigation and adaptation constitutes a foundational framework for effectively confronting climate change. Mitigation centres on curbing greenhouse gas emissions and reining in anthropogenic activities that drive global temperature rises; adaptation, conversely, seeks to attenuate climate impacts and cultivate resilience to phenomena that are already unfolding. The relationship between these strategies, however, is never linear and demands continuous negotiation.

At its most fundamental, this interplay demands careful choreography between long-lived structural transformations and the flexible, anticipatory measures demanded by an already-dynamic climate. Mitigation targets the symptomatic drivers of climate change through regulatory frameworks, emerging technologies, and shifts in societal practice. By contrast, adaptation acknowledges the reality of climate impacts—drifting climatic averages, enhanced frequency of extreme events, and reconfiguring biological systems—by mobilising measures that reduce vulnerability and safeguard human and ecological systems.

Central to their effective integration is the imperative for holistic planning that incorporates both strands from the outset. Synergistic outcomes emerge only when decision-makers evaluate mitigation and adaptation simultaneously, recognising the reciprocal influences. The construction of urban green spaces, for example, serves a dual purpose: it lowers the urban heat footprint (a mitigation gain) while simultaneously elevating community preparedness for heatwaves and flooding (an adaptation dividend).

Crucially, any effective climate response must identify and weigh the trade-offs and synergies created when mitigation and adaptation actions intersect. Policymakers, practitioners, and affected communities require a detailed grasp of these relationships to identify pathways that yield the highest possible co-benefit while containing any adverse side effects.

The mutual reliance of mitigation and adaptation reinforces the necessity of moving beyond siloed thinking. Governance that genuinely links these two strands must involve a wide spectrum of actors: local and national authorities, the private sector, non-governmental organisations, and community groups. By nurturing collaborative networks and open exchanges of knowledge, we can co-create solutions that respect the local context, satisfy short-term adaptive pressures, and remain robust over the long horizon of sustainability.

Ultimately, attention to the interplay of mitigation and adaptation must remain both sophisticated and strategically coordinated. Faced with the layered complexity of climate change, we cannot afford to treat these approaches in isolation. A well-informed and integrated response is the only means to produce resilient pathways that protect both society and the ecological systems on which it relies.

Evaluating Cost-Benefit Analyses

Cost-benefit analyses serve as indispensable tools in the formulation of climate mitigation and adaptation measures. Through these evaluations, policymakers are enabled to juxtapose prospective expenditures and gains that accompany differing interventions and policy frameworks. Because the ramifications of climate change are characterised by temporal disaggregation and probabilistic uncertainty, appraising the cost-efficiency of any given measure becomes particularly intricate. Nonetheless, a methodical disaggregation of costs and benefits yields guidance on the optimal configuration of resource deployment, thereby enhancing aggregate societal welfare.

In undertaking cost-benefit evaluations of climate-related initiatives, it is incumbent to integrate both quantifiable

CLIMATE CHANGE IN AMERICA

and non-quantifiable dimensions. Quantifiable expenditures encompass capital outlays, recurrent operational disbursements, and upkeep costs tied to the construction of infrastructures and the deployment of technologies. Conversely, non-quantifiable benefits—including gains to public health, the maintenance of biophysical systems, and the fortification of social cohesion—, while resistant to numeric encapsulation, possess substantial societal importance. Accurate incorporation of these intangible gains is therefore essential for a nuanced comprehension of the net impact of climate interventions.

Moreover, the selection of an appropriate social discount rate to recalibrate future costs and benefits remains a central consideration in climate-related cost-benefit analyses.

Discounting permits comparison of temporal costs and benefits, embodying society's temporal preferences. The choice of discount rate, however, remains contentious, intertwining questions of intergenerational justice with the ethical valuation of future lives. Further complexity arises in the treatment of uncertainties linked to climate projections and the efficacy of both mitigation and adaptation strategies. Systematic sensitivity analyses complemented by scenario approaches expose the dispersion of possible futures and highlight options robust to differing trajectories. Transparency in articulating underlying assumptions and the spectrum of uncertainties is indispensable for earning the trust of stakeholders and for the credibility of the analysis. Explicitly conveying the uncertainties and methodological constraints to decision-makers fosters a more nuanced and informed deliberation. The strongest justification for climate investment emerges when the adjusted benefits demonstrably eclipse the initial expenditures. By fortifying the empirical foundation and clarifying the underlying trade-offs, cost-benefit frameworks not only elucidate the economic rationale for action but also inform the design of policies that are both resilient to future change and equitable across generations.

These analyses function as critical instruments for managing the intricate processes involved in climate change mitigation and adaptation. By providing a rigorous examination of vulnerabilities and response strategies, they facilitate the prudent allocation of resources, thereby protecting the livelihoods of present populations while securing the environmental and economic foundations for generations to come.

Innovative Technologies in Climate Strategy

The advancement and deployment of cutting-edge technologies are not just significant; they are also transformative and inspiring to the trajectory of global climate strategy. Over the last decade, these transformative breakthroughs spanning multiple sectors have furnished viable pathways for dampening the trajectory of climate change. Within the energy domain, sustained progress in solar, wind, and hydropower technologies has lowered capital costs and increased conversion efficiencies, sharply curbing both emissions and dependence on fossil fuels, parallel to declining marginal costs. Complementary advances in energy storage—ranging from lithium-ion to emerging solid-state and flow battery chemistries—are adding the grid-scale capacity and mobility essential for a resilient, renewable-dominant electricity system.

Smart grid architectures, a crucial layer in our climate strategy, are revolutionising electricity distribution and demand management. By integrating Internet of Things devices and leveraging machine-learning algorithms, grid operators can monitor, predict, and fine-tune electricity flows across millions of nodes. This flattens load peaks, integrates distributed energy resources, and reduces curtailment losses, thereby enhancing the resilience of our energy systems.

In the mobility domain, electrified transportation is a game-changing innovation. Battery-electric buses, two-wheelers, and passenger cars are replacing diesel and petrol combustion, and ongoing innovations in battery chemistries promise to further enhance energy density and charging speed. Moreover, vehicle-to-everything (V2X) communication and autonomous driving algorithms are set to revolutionise the industry. These technologies will harmonise charging loads with grid demands and renewable output, while orchestrating platooned flows that reduce drag and emissions across entire city corridors.

Recent developments in agricultural technology have strengthened climate adaptation efforts. Data-intensive precision farming—encompassing geospatial crop analytics and variable-rate irrigation—improves water-use efficiency and lowers nitrous oxide emissions. Concurrently, modern biotechnologies have produced genetically engineered cultivars exhibiting increased tolerance to drought and heightened resistance to key pests, thereby fortifying food systems against climatic volatility while being compatible with lower fertiliser inputs.

The convergence of climate science and computational technology has enriched climate modelling and impact forecasting. Analyses conducted on next-generation supercomputers deliver finer-resolution projections of heatwaves, floods, and cyclones, enhancing preparedness at national and municipal levels. Complementary to these efforts, remote sensing platforms—satellites and Light Detection and Ranging instruments—generate timely, synoptic datasets on land-surface phenology, soil moisture, and forest-stored carbon, thereby enabling adaptive management of critical ecosystems.

Looking ahead, novel carbon management strategies, such as carbon capture and storage, offer pathways for curtailing emissions from fossil-fuel and industrial processes. By separating carbon dioxide at the point of emission and transporting it to geologically stable formations for long-term storage, CCS can materially lower the trajectory of atmospheric greenhouse-gas concentrations.

In sum, marrying cutting-edge technologies with climate strategy creates numerous avenues for confronting environmental pressures. Success hinges on continuous dialogue among businesses, research institutes, and governing bodies that can accelerate technological development and facilitate the broad uptake of these tools, ultimately steering society toward a more sustainable trajectory.

Supportive Policy Frameworks for Equitable Choices

Concurrently advancing mitigation and adaptation demands a policy architecture that not only encourages cohesive, synergistic responses to the climate challenge but also empowers enterprises, neighbourhoods, and individuals to embrace sustainable trajectories. Decision-makers at the national, regional, and municipal tiers must structure the legal and financial milieu that empowers these entities to make responsible and sustainable choices.

A central pillar of sound policy design consists of persuasive, well-articulated climate objectives. Formulating explicit, science-based quantitative targets for emission curtailment and for bolstering resilience provides a practical navigation chart and injects urgency into the diverse constituencies involved. Such targets also furnish the empirical and normative foundations for fashioning regulations, performance standards, and fiscal inducements that can trigger research, nurture market-ready solutions, and mobilise capital toward the emergence of low-carbon technologies.

Alongside the establishment of targets, climate policy frameworks must also embed flexibility and adaptability as core principles. The dynamic and uncertain character of climate change mandates that policies can be recalibrated in light of emerging scientific insights, breakthroughs in technology, and shifting socio-economic dynamics. This adaptability promotes the capacity to refine and redirect both mitigation and adaptation initiatives, thereby safeguarding their effectiveness over the long duration necessary to avert global climate thresholds.

Equally important is the mainstreaming of climate considerations across all relevant sectors. Integrating climate objectives with energy security, public health, economic development, and social equity, among others, ensures that efforts are mutually reinforcing rather than counterproductive. By embedding climate priorities within each policy domain, governments can identify and amplify synergies while mitigating potential conflicts, thereby achieving more coherent and durable sustainability outcomes.

Equity and inclusiveness must be at the forefront of our policy design. Socio-economically marginalised and vulnerable populations often bear the brunt of climate consequences while having the fewest resources to respond. Therefore, our policy designs must prioritise the identification and fulfilment of their needs through directed investment, adaptive support schemes, and inclusive, transparent decision-making. Strengthening resilience at the community level not only reduces exposure to future shocks but also fosters social solidarity and empowerment.

Furthermore, successful policy design hinges on well-defined coordination mechanisms that link public agencies, private sector actors, civil society groups, and international collaborators. Effective partnerships and the open exchange of data diminish the likelihood of overlapping efforts, optimise the allocation of financial and human resources, and capitalise on sector-specific knowledge. Institutionalising regular forums for dialogue and co-development enables actors to disseminate successful practices and to design ground-breaking, context-responsive strategies collectively.

Equally, systematic monitoring, regular reporting, and open communication constitute the bedrock of any credible policy package. Rigorous monitoring and evaluation frameworks empower governments to measure advancement, pinpoint shortfalls, and ensure that all parties fulfil their obligations. Transparent disclosure of data and policy intent keeps the public apprised of strategic decisions, investment prospects, and emerging vulnerabilities, thereby reinforcing public confidence and mobilising broader societal backing for climate initiatives.

In summary, policy frameworks that harmonise climate mitigation and adaptation efforts must embody ambition, pragmatic flexibility, horizontal integration, and equitable distribution of costs and benefits. By ensuring that the costs and benefits of climate policy are distributed fairly, legislators and administrators can make sustainability advocates feel that their efforts are making a real difference in the fight against climate change.

Case Studies: Successful Integrated Models

This section presents selected case studies demonstrating effective integrated frameworks that harmonise mitigation and adaptation actions to address the effects of climate change.

Case study 1: Portland, Oregon's Climate Action Plan. Portland's climate action framework embodies an integrated response that weaves mitigation and adaptation into a unified trajectory. The city has advanced a series of mitigation initiatives, including transit-oriented development, energy-efficient retrofits, and aggressive investments in renewable energy. Simultaneously, Portland has elevated resilience by expanding green infrastructure networks, upgrading flood management systems, and embedding climatic variability into every layer of urban planning. The success of this combined approach lies in its ability to tackle climate hazards while embedding sustainability into the city's growth DNA.

Case study 2: The Netherlands' Delta Programme Confronted by the persistent threats of sea-level rise and increased flooding, the Netherlands has developed the Delta Programme as its long-term, integrated climate risk management framework. The Programme weaves together elevated coastal defences, forward-looking land-use planning, and cutting-edge water management techniques to reduce flood risk and to strengthen adaptive capacity across the delta. By coordinating spatial, hydraulic, and infrastructure investments, the Delta Programme exemplifies how a small, low-lying nation can methodically harmonise mitigation of carbon emissions with preparation for irrevocable climatic shifts. By marrying engineering prowess with ecological insight, the Delta Programme exemplifies how an even-handed strategy can successfully shield societies from the escalating consequences of a warming world.

Case study 3: Singapore's Sustainable Development Blueprint Singapore, globally acknowledged for its sustainability fervour, has formulated a proactive roadmap that fuses mitigation and adaptation within its Sustainable Development Blueprint. This all-encompassing strategy pursues immediate carbon reduction, superior energy efficiency, and a gradual shift to renewable sources, while concurrently preparing for higher sea levels, intensified weather extremes, and the burgeoning urban heat island. By embedding both climate mitigation and adaptation in a cohesive developmental narrative, Singapore illustrates how an integrative design can foster lasting ecological and societal resilience. Together, these examples highlight the value of an approach that deliberately synchronises carbon and climate adaptation, showing that such consolidated frameworks can tackle the complex, interlocking challenges that climate change presents.

Challenges in Implementation

Successfully implementing effective climate mitigation and adaptation strategies necessitates confronting a range of interrelated challenges, each of which must be tackled for climate action plans to produce meaningful outcomes. Central to the difficulty is the multidisciplinary character of the required interventions, which demand sustained cooperation among a wide array of actors that include governmental agencies, private firms, civil society organisations, and local communities. While collaborative frameworks exist, the practical coordination of diverse objectives, practices, and capacity levels across these groups frequently proves to be a protracted and complex endeavour. In parallel, the scarcity of financial and human resources, particularly in developing countries and within marginalised populations, restricts the feasibility of executing wide-ranging climate measures. Addressing this limitation necessitates the design of innovative, scalable financing instruments, as well as frameworks that assure the just and effective allocation of the available resources. Compounding these logistical considerations is the political environment in which climate policies operate; securing sustained public backing for transformative measures often requires skilful engagement and transparent, evidence-driven communication to counter scepticism and mobilise constituencies. The inherently evolving character of climate risks further complicates planning, as stakeholders must weigh multiple future scenarios that feature high levels of uncertainty. Consequently, fostering institutional adaptability and resilience becomes a non-negotiable condition of successful implementation. Finally, existing legal and regulatory constraints can obstruct timely action; reviewing and, when necessary, recalibrating statutory and regulatory instruments is indispensable for allowing climate strategies to unfold without unnecessary delays or contradictions.

Cultural and behavioural shifts are also necessary to ensure that sustainable technologies and practices are embraced at scale. In addition, the absence of rigorous monitoring, evaluation, and accountability mechanisms compromises the ability to measure the actual impact of any measure enacted. Tackling these barriers demands an integrative strategy that weaves together technological advancement, coherent policymaking, active societal participation, and transnational cooperation. Once we recognise these impediments and confront them methodically, we lay the groundwork for climate mitigation and adaptation that is not only efficient but also meaningful, thereby securing a resilient and sustainable world for future generations.

Conclusions: Toward a Sustainable Future

Reflecting on the mitigation and adaptation pathways we have examined, it is clear that the journey to a sustainable future is strewn with intricate challenges. Effective climate action must marry these pathways to confront the diverse dimensions of climate risk. Each strategy that we deploy comes with implementation barriers, yet together they hold the promise of genuine and lasting change. Realising that promise, however, is contingent on an all-encompassing orientation that combines decisive political commitment, informed public participation, cutting-edge innovation, and a sturdy policy backbone. Only through such multidimensional effort can we chart a future that is both liveable and equitable.

A central insight from our deliberations is the necessity of reorienting our entire approach to climate change. Mitigation and adaptation cannot continue to occupy isolated silos; they ought to be integrated strands woven into a single strategic framework. Achieving this integration demands that we prioritise the cultivation of resilience through infrastructure and capability investments, while simultaneously initiating steep, verifiable reductions in greenhouse gas emissions. It also implies a structural rethinking of our economic and social architectures, such that sustainability becomes the governing criterion at every level of policy and practice.

Illustrative case studies throughout the process have documented instances in which municipalities, corporations, and national authorities have pursued combined mitigation-adaptation pathways with demonstrable results. These narratives offer concrete models of a future in which sustainability is manifest rather than aspirational. Their successes point to the indispensable roles of cross-sectoral partnerships, the systematic exchange of expertise, and the engagement and agency of local populations in engineering transformative outcomes.

A commitment to sustainability requires continued collective engagement anchored in shared responsibility. The time for decisive action is now, and success depends on every citizen, every organisation, and every government participating meaningfully. We must now direct capital toward clean technologies, promote cross-sector partnerships, and embed stewardship of the earth into every decision. Only by maintaining shared purpose and collective action will we chart a course towards a planet in which vibrant ecosystems, equitable economies, and a secure climate are not only equitable but self-sustaining for generations to come.

Reflecting on the delicate interplay of mitigation and adaptation, we discern a matrix of hurdles and openings that climate action continually erects and reframes. The task may appear forbidding, yet its contours are dense with transformative potential. When we bind our endeavours to an unwavering purpose, we do not merely defend the planet; we redesign it. The reward of persistence is a tomorrow that is beautiful, just, and rich in the resources that life requires.

10
Future Directions
Navigating Political and Social Complexities

Current Political Landscape: Consequences for Climate Policy

The present political environment consists of an interrelated set of ideologies and strategies that directly determine the velocity and scope of climate policy and practice. The pendulum of political authority that swings between competing governance philosophies now translates, with consistent regularity, into movements—both constructive and disruptive—within statutes, funding allocations, and regulatory frameworks. Hence, a systematic evaluation of these movements and the systemic conditions that generate them is essential for anticipating the future configuration of climate advocacy and operational responses. Discourse on climate change persists across a spectrum of political positions, each embedded in distinct belief systems, stakeholder preferences, and accountability pressures. Attention to the rhythm of political speech, the sequencing of policy signals, and the material consequences of governance choices therefore yields rich descriptive and predictive data concerning the viability of environmental integrity at the inter-jurisdictional, national, and global scales. The same investigation further reveals the agents and arrangements that front the political theatre, from elected leaders and bureaucratic coalitions to lobby groups and multilateral institutions. The preferences and tactical choices of these central actors, in their budget negotiations and regulatory orientations, establish non-reversible near-term conditions that determine whether climate precautions evolve into binding, scalable practice.

Equally influential is the mapping of public attitudes alongside the congruence or divergence of these attitudes from prevailing political positions on environmental matters. The reciprocity between societal sentiment and political resolve creates the latent energy that propels climate advocacy, thereby revealing the critical need for systematic and inclusive engagement with the spectrum of community perspectives. Concurrently, the nexus of government, corporate actors, and organised interest groups constitutes a vital stratum of the political ecology that governs climate policy. The negotiations arising within these arenas frequently determine the trajectory of regulatory choices and the distribution of financial and institutional resources, accentuating the complex, co-constitutive linkage between electoral calculus and ecological necessity. A granular appraisal of the prevailing political tableau thus yields a calibrated comprehension of both the barriers that must be surmounted and the strategic openings that may be prudently leveraged in the quest for effective climate advocacy.

Identifying Key Stakeholders in Climate Advocacy

Strategically mapping key stakeholders in climate advocacy is essential for orchestrating effective, long-term climate action. The stakeholder constellation encompasses government agencies, non-governmental organisations, research institutions, the private sector, and civic collectives.

Each constituency brings distinctive capabilities and leverage points that collectively advance the climate agenda. Regulatory agencies possess the mandate to craft and enforce legislation that governs emissions and resource use. Non-governmental organisations mobilise data, networks, and narrative to refocus political will and galvanise public engagement. Research institutions generate peer-reviewed knowledge and incubate low-carbon technologies, while also training the next generation of climate-informed leaders. The private sector, especially in resource-intensive industries, is poised to shift supply chains and finance cleaner alternatives through capital investment and innovation. Finally, civic collectives ground advocacy in lived experience and tailor solutions to the socio-economic fabrics of neighbourhoods, thereby embedding climate justice into every initiative.

Legislative Challenges and Opportunities

The arena of climate legislation constitutes a multidimensional space influenced by a broad array of stakeholder interests, competing political currents, and the varying exigencies of the public. In considering the legislative trajectories ahead, it becomes evident that the formulation of durable and effective climate policies is far from straightforward. One of the primary impediments is the necessity of forging a coherent consensus among legislators whose priorities, electoral bases, and ideological orientations differ markedly. The imperative for prompt climate action must be communicated in a manner that underscores its immediacy, while not

CLIMATE CHANGE IN AMERICA 159

attenuating attention to concurrently urgent matters such as health care, public safety, and social equity.

In addition, the legislative journey toward substantively ambitious climate measures is often complicated by the mobilisation of industry lobbyists and well-resourced interest factions whose analytical and strategic capacities rival those of public agencies. Their engagement in the policymaking arena can skew the framing of debates and the formulation of regulatory details. The persistent contention concerns the equitable reconciliation of ecological integrity with the legitimacy of economic expansion and the preservation of employment. Crafting politically viable solutions thus requires the construction of policy architectures that integrate emission reduction trajectories with compensatory mechanisms, transitional assistance, and innovation incentives for jurisdictions and workforces whose current livelihoods hinge on carbon-intensive practices.

At the same time, the widening recognition that climate change can serve as a catalyst for innovation, workforce diversification, and technological progress presents a strategic opening. When policy advocates engage lawmakers with a clear, actionable articulation of these advantages, they can frame climate measures not as burdens but as vehicles for economic reinvigoration, thereby inspiring creative, forward-thinking legislation.

Meaningful progress within the labyrinthine policy arena depends on a nuanced reading of the prevailing political climate. Fluctuating power distributions, pronounced partisan fissures, and the rhythmic pressures of electoral calendars jointly delineate the contours of legislative possibilities. Effectively mapping these variables to pinpoint fleeting opportunities for cross-party collaboration remains vital to strengthening durable climate measures.

Maximising periods when scientific agreement, public awareness, and institutional readiness converge requires policy architects to build instruments that mirror prevailing public attitudes. When regulations authentically reflect communal values, their design, adoption, and enforcement enjoy broader legitimacy, enabling them to penetrate and mobilise disparate constituencies.

In sum, the pursuit of durable climate legislation demands a calibrated integration of realistic appraisal, inventive reasoning, and principled compassion. By confronting historical resistances while simultaneously advancing anticipatory proposals, advocates create the practical, fair, and implementable policy frameworks that the climate crisis urgently demands.

Public Opinion: Shaping Climate Policy

Public opinion occupies a fundamental position in the formulation of climate policy, serving as a barometer of societal concern and commitment to environmental challenges. Given the accelerating evidence of climate-related threats, policymakers must attend closely to populist feelings to design durable and politically viable responses. A systematic examination of the drivers of public opinion on climate change reveals a complex interplay of media exposure, socioeconomic status, cultural values, and educational attainment.

Media institutions function as the primary channel through which climate knowledge reaches citizens. The manner in which environmental stories are framed can decisively alter public understanding and, consequently, the political appetite for policy action. When journalists foreground scientific consensus and document tangible climate impacts, they cultivate a more informed and potentially mobilised electorate. The recent shift toward digital platforms has further diversified the information landscape, enabling activist networks and peer-to-peer communication to contest, reinforce, or redefine the dominant climate narratives encountered by the public.

Socioeconomic dimensions fundamentally shape public attitudes on climate policy. Variations in income, access to technology, vulnerability to climate impacts, and general financial security collectively determine how urgently different groups prioritise climate action. Recognising these factors enables the development of climate strategies that are both fair and effective, ensuring that the most affected populations receive commensurate attention and that the broader society advances toward justice-centred solutions.

At the same time, cultural norms and educational experiences nuance public sentiment. Deeply embedded values regarding nature, resource use, and community health influence whether individuals regard climate protection as a moral imperative or a distant concern. Robust educational experiences—especially those that emphasise scientific literacy, critical appraisal of information, and interdisciplinary perspectives on sustainability—equip citizens to participate substantively in policy debates and to champion measures that are grounded in rigorous evidence.

To accurately measure public opinion, researchers must employ a deliberate combination of quantitative and qualitative techniques: structured surveys, deliberative focus groups, and in-depth narrative analyses, for instance, are indispensable for capturing the range of lived experiences and cognitive frames that inform attitudes. Policymakers who attend to the complex layering of these views can shape climate programmes that not only address empirical risk but also resonate meaningfully with diverse values, thereby galvanising broad, sustained commitment to a durable and equitable climate future.

Media's Influence on Climate Dialogue

Media institutions function as key intermediaries between scientific research on climate change and public comprehension, thereby modulating public attitudes and policy trajectories. Their capacity to curate, frame, and circulate information confers considerable power over whether climate change is perceived as a distant, abstract threat or as an immediate, actionable exigency. The manner in which scientific consensus, regulatory initiatives, and local manifestations of climate impacts are presented to the constituency can, in turn, steer civic engagement and policy advocacy. For these reasons, the journalistic mandate requires that reporting be accurate, nuanced, and sufficiently analytical so that citizens can engage in reasoned deliberation and so that public opinion can evolve in a fact-informed manner.

Beyond bare reporting of data and legislative milestones, the journalistic vocation involves a commitment to unpack and contextualise climate phenomena that causally interlace with a spectrum of economic, social, and ecological dimensions. Through sustained investigative projects, multi-scalar reporting, and interdisciplinary collaboration, media practitioners can illuminate the reciprocal dependencies between fossil fuel economies, patterns of vulnerability, environmental justice, and the policymaking process. This approach not only enriches public understanding of the climate crisis as a complex, non-linear process but also enhances the capacity of community stakeholders and decision-makers to identify effective and equitable adaptive strategies.

Further, the media possesses the capacity to magnify the circumstances of communities that climate change routinely marginalises. By prioritising testimonies and viewpoints from these groups, journalism situates the human toll of climate phenomena at the forefront, encouraging wider empathy, greater awareness, and unified responses. Responsible reportage can seamlessly weave empirical evidence with lived experience, permitting a public conversation about climate change that is both scientifically accurate and emotionally resonant.

Beyond informing and narrating, the media also functions

as a crucible for public advocacy and involvement. Through a diverse array of formats—be it investigative documentaries, pointed op-eds, or interactive online modules—journalism can galvanise viewers, readers, and listeners to engage with climate initiatives, support progressive legislation, and demand accountability from leaders. By maintaining a vigorous realm for debate, the media can unsettle entrenched myths, provoke analytical questioning, and cultivate a citizenry that is both well-informed and prepared to act.

Yet, the evolving media environment poses significant obstacles to the preservation of objectivity and evidential fidelity in climate discussions. Misinformation, sensational framing, and lobbying-driven communications frequently cloud the scientific consensus and impede the formulation of sound, evidence-based policy. Media practitioners are therefore charged with the obligation to adhere to high ethical norms, subject every source to thorough verification, and consistently differentiate rigorous scientific inquiry from speculative or politically motivated assertions.

In the face of climate change's multilayered challenges, the media retains its essential role as an architect of public comprehension, civic involvement, and decisive action. By informing, educating, and galvanising communal responses, the media cultivates the conditions from which transformative, collective action may emerge, illustrating the power of disciplined, ethical climate journalism to shape a resilient and informed society.

Economic Considerations and Industry Impact

Climate change and economic frameworks are now inseparable, requiring sophisticated analysis to understand how the shift to a sustainable, low-carbon economy reverberates through global industries. Recognition of the climate emergency has elevated the financial risks and opportunities associated with decarbonisation, making the economic dimensions of the transition a focal concern for policymakers and corporate leaders alike. The imperative now is to map the economic landscape, identify potential disruptions, and design mechanisms that not only buffer against shocks but also harness the growth potential of a greener economy, offering a hopeful vision for the future.

Central to these economic analyses is the interplay between ecological viability and financial resilience. Corporations in every segment of the economy are reassessing value chains, governance structures, and product lifecycles to align with rising expectations for environmental stewardship. This recalibration often spurs technological advancement and operational efficiencies but can also strain balance sheets and demand sophisticated capital management. Strategic resource allocation, therefore, must balance short-term cost pressures with the longer horizon required to internalise the costs of carbon and other environmental externalities.

The shift to renewable energy and low-emission technologies also generates a distinctive set of economic opportunities. Capital is increasingly directed towards green infrastructure and circular-business models, with investors and boards recognising that future profitability will be tightly linked to sustainability performance. Supportive governmental measures—ranging from tax incentives to coordinated public-private financing mechanisms—amplify these incentives, underscoring how well-designed policy can steer private capital and innovation towards climate-aligned outcomes. Thus, the economic calculus of the transition is not only about risk mitigation but also about positioning for competitive advantage in an evolving, climate-constrained global marketplace.

The transition from carbon-intensive energy modalities and associated production paradigms compels analysts to scrutinise the full market cascade that such a shift provokes. Regions and labour markets that have historically depended on fossil-fuel value chains are experiencing structural realignment; these communities will require targeted safeguards and coordinated transition strategies that reconcile decarbonisation targets with socio-economic stability. The confluence of climate imperatives and livelihoods, therefore, mandates that regulators adopt an integrated analytical lens throughout the lifecycle of policy design and execution.

Simultaneously, the interwoven fabric of the global economy stipulates that the financial fallout of climate action is, by definition, cross-border. The reconfiguration of supply webs, the recalibration of commercial treaties, and the redefinition of geopolitical alliances are all being reframed by climate risk, underscoring the necessity for a multilateral governance architecture. Diplomatic fluency, underwritten by a granular grasp of these dynamic cross-sectional impacts, becomes a prerequisite for orchestrating a coherent, unified global stance on climate perturbations. This underscores the vital role of global collaboration in addressing climate change.

At a foundational level, a comprehensive understanding of the economic and industrial ramifications of climate policy is an indispensable precursor to strategic decision-making. By ensuring that debates encompass business imperatives, governance vectors, and broad societal welfare, analysts affirm that economic rationality must be married to environmentally responsible stewardship at every tier of implementation.

The Influence of Grassroots Movements

Grassroots movements have been crucial in redefining public discourse and policy on climate change. Fuelled by committed individuals and locally anchored organisations, they have succeeded in cultivating heightened awareness, galvanising broad public backing, and exerting sustained pressure on lawmakers to elevate climate matters on policy agendas. Such initiatives are anchored in the conviction that lasting change germinates within communities, where neighbours band together to pursue shared climate-related goals.

A principal asset of these bottom-up campaigns is their capacity to connect the lived experiences of communities to the desks of policymakers. By elevating a mosaic of voices that often remain unheard in elite circles, they compel the formal policy arena to reckon with the disproportionate harms that climate change inflicts on the most vulnerable populations. Campaigners have, in numerous instances, succeeded in embedding considerations of environmental justice and equity into legislative frameworks, thus challenging the long-standing marginalisation of groups that have long borne the brunt of environmental degradation and policy indifference.

Grassroots initiatives remain essential in fostering behavioural transformations and advancing sustainable practices within local populations. Through curricula, workshops, and persistent outreach, these groups enable citizens to integrate eco-conscious routines, shrink their carbon footprints, and enlist in neighbourhood conservation programmes. When these dispersed, single actions coalesce at the grassroots level, their cumulative weight can produce quantifiable ecological gains and advance expansive sustainability objectives.

Creativity and experimentation lie at the heart of grassroots movements. Small communities frequently pilot alternative technologies, renewable energy co-ops, and site-specific responses to ecological pressures. The successful demonstration of such localised experiments frequently persuades higher-tier policymakers and industry to adopt greener, more sustainable frameworks, thus spreading their original benefits beyond the communities that conceived them.

The emergence of social media and digital networks has further magnified the effectiveness of grassroots efforts. Online channels enable real-time information exchange, synchronised advocacy, and the worldwide amplification of locally driven actions, linking isolated issues to the shared global climate agenda. This digital openness has fostered transnational networks of climate-focused citizens, who, albeit dispersed, unify in shared purpose and mutual support, multiplying the influence of their localised endeavours.

In summary, grassroots movements exert a decisive influence on contemporary climate change discourse. Through advocacy, public education, technological innovation, and community mobilisation, they not only supplement but occasionally recalibrate official policy arenas. These movements have crystallised into a palpable, multidimensional force that reorients climate governance towards more participatory and locally attuned outcomes. As the global community grapples with the nonlinear and uncertain dimensions of climate phenomena, a rigorous appreciation of grassroots dynamics will be essential. Only then can policy formulations be truly inclusive, adaptive, and resistant to the fragile contingencies that the climate crisis entails.

Navigating Partisan Divisions

Partisan divisions over climate policy present a daunting

obstacle that demands both tactical political insight and the capacity to traverse ideological boundaries. In the United States, climate change has often been ensnared in heightened polarisation, marked by divergent views of the problem's magnitude, the legitimacy of proposed responses, and the extent to which federal governance ought to be involved. A rigorous examination of the origins of this bifurcation is essential for crafting policy responses that are both durable and politically viable. Among the principal drivers of partisan fracture are several interrelated elements.

Foremost is the divergent political philosophy surrounding the state's role in environmental governance: conservative thought typically prizes market-led remedies and a circumscribed public role, whereas progressivism more readily embraces comprehensive regulation and state intervention. Entrenched interest blocs further complicate the calculus; the fossil fuel sector, for example, exerts considerable leverage over one political faction, while environmental NGOs often mobilise the opposing camp. These constituencies do not merely lobby in a vacuum; they actively reshape party platforms and conversational frameworks regarding climate. To complicate matters, the contemporary media ecosystem, stratified along partisan lines and fortified by algorithmic reinforcement, magnifies and perpetuates the polarisation, rendering cross-cutting dialogue increasingly rare.

To effectively address these obstacles, the imperative is to identify overlapping interests and cultivate inclusive processes that honour the varied apprehensions and priorities of all stakeholders. One productive strategy is to recast climate change as a shared concern by foregrounding universally respected values such as public health, national security, and equitable economic advancement. Cross-ideological coalitions, bolstered by structured and respectful conversations among stakeholders who hold divergent convictions, can further reduce the corrosive effects of partisan polarisation.

Highlighting local consequences and community-driven resilience initiatives helps to recentre the debate on actionable, apolitical, and observable outcomes. Leadership that rises above partisan calculus, intentionally choosing the public interest over electoral gain, remains a linchpin of these efforts. By cultivating cooperative arrangements and a communal ethic of responsibility, society can attenuate the corrosive effects of division and advance substantive climate mitigation. Ultimately, the durability of climate policy will hinge on a steadfast dedication to overlapping concerns, the reconciliation of differences, and a governance ethos anchored in cooperative compromise.

Innovative Approaches in Policy Formulation

In light of growing insights into the intricate dimensions of climate change, the demand for innovative policy approaches has grown urgent. Conventional approaches often falter when confronted with the multidimensional character of climate phenomena, struggle to integrate varied stakeholder viewpoints, and lag behind the rapid advancement of scientific knowledge. Under these conditions, the deliberate adoption of innovative policy-design practices becomes essential for triggering substantive change and for fortifying the resilience of human and ecological systems.

A strategic starting point for innovation lies in the systematic assembly of interdisciplinary expertise at every stage of policy development. When scientists, economists, social scientists, policymakers, and community representatives collaborate from the outset, the resulting policies are more likely to capture the intricate relationships among the risks, opportunities, and social considerations climate change presents. This integrative process not only grounds policy in rigorous, evidence-based reasoning but also embeds considerations of inclusion and equity into the very structure of the policy, fostering legitimacy and broad acceptance.

In parallel, the deliberate application of emerging technologies holds the potential to transform the policy formulation landscape. Sophisticated modelling techniques, robust data analytics frameworks, and interactive visualisation instruments empower decision-makers to project alternate future pathways, evaluate the contingent benefits and trade-offs of proposed interventions, and articulate complex scientific insights in accessible terms. Digital platforms further extend the reach of the policy discourse, enabling a wider array of citizens, activists, and local experts to contribute observations and suggestions that enrich the policy debate and ensure that a broader spectrum of lived experiences informs the final decisions.

Adaptive governance frameworks offer a promising paradigm for crafting policies under conditions of uncertain climate futures. Such frameworks foreground flexibility, iterative responsiveness, and the continual incorporation of new information, acknowledging that climate-related stressors will keep shifting. By embedding iterative review processes and conditions for recalibration, decision-makers can tune policies in light of observational data, new risk vectors, and sometimes unexpected perturbations, thus bolstering the robustness and elasticity of the policy suite.

Cooperative public-private alliances further enrich policy innovation by amalgamating different pools of capital, technical know-how, and stakeholder motivations around common climate goals. These collaborations can catalyse breakthroughs in blended-finance models, accelerate the deployment of low-carbon technologies, and streamline operational implementation, thereby embedding sustainability into the policy architecture itself.

The effective uptake of such novel methods, however, requires a collective embrace of a horizon-scanning, experimental ethos and unwavering cross-institutional cooperation. As the policy community steers through the intricate and interactive layers of climate governance, a continual search for and investment in unconventional tools will prove decisive in engineering a resilient, adaptive socio-ecological future for both present and succeeding generations.

Vision for a Collaborative Future

When we shape our vision for a collaborative future, we must first appreciate how intimately our global communities are linked in the fight against climate change. Looking forward, the vision we must nurture emphasises coordinated, multilayered action that crosses every geographical, political, and socio-economic divide. The animating principle of this vision is the alignment of a broad spectrum of actors—governments, the private sector, non-governmental organisations, and every concerned citizen—around a common pledge to reduce emissions and to embed sustainable practices in every facet of society.

Central to this collaborative future is the deepening of international partnerships. Because the earth's climate system knows no borders, countries are compelled to adopt open, continuous dialogue and to embark on joint projects that unravel the complex, interconnected dilemmas we face. Gathered in international summits, diplomatic tracks, and dynamic networks that circulate research and innovation, states can exchange their distinctive capacities—financial resources, advanced technology, and indigenous ecological knowledge—thereby magnifying the overall efficacy of global climate action.

The vision of a future anchored in collaboration places inclusive and equitable participation at its core. It insists on the necessity of integrating the knowledge and lived experiences of indigenous communities, historically marginalised groups, and other vulnerable populations in all stages of decision-making. When diverse voices are not only heard but actively prioritised, collaborative initiatives can produce responses that are genuinely attuned to the distinct needs and lived realities of people around the world. Such inclusivity enriches the policy development process and simultaneously cultivates a deep sense of ownership and empowerment among communities that face the direct consequences of climate change.

Technological progress will be instrumental in translating this collaborative vision into practice. Cutting-edge tools and digital platforms can enable instant data sharing, refine predictive modelling, and offer cost-effective options, thereby strengthening synchronised responses to climate-related threats. The integration of artificial intelligence, blockchain, and geospatial mapping can further optimise the distribution of resources, improve disaster preparedness, and facilitate evidence-based decision-making at both local and global scales.

In addition, the prospect of a collaborative future recognises public-private partnerships as engine rooms of lasting transformation. By intelligently linking government, business, and civil society, such partnerships marshal capital, know-how, and inventive capability into projects that are both scalable and sustainable. The synergies created can stimulate inclusive economic advancement, accelerate the diffusion of breakthrough technologies, and craft solutions that judiciously cross and converge traditional industry lines.

Ultimately, the collaborative future rests on a relentless pledge by individuals to embrace sustainable living and to champion systemic change wherever they operate. Choosing low-impact products, backing enterprises that prioritise the planet, and pressing for progressive policy reform are, individually and together, decisive contributions to the emergence of a low-carbon, circular economy. When neighbourhoods are animated, stewards are cultivated, and shared responsibility is reinforced, citizens transmute passive expectation into active participation, thereby co-authoring a durable and equitable future for future generations.

References For Further Reading

Chapter 1: Introduction - Understanding the Climate Crisis in America

- **IPCC, 2021:** Climate Change 2021: The Physical Science Basis. Contribution of Working Group I to the Sixth Assessment Report of the Intergovernmental Panel on Climate Change [Masson-Delmotte, V., P. Zhai, A. Pirani, S. L. Connors, C. L. Péan, S. Berger, N. Caud, Y. Chen, L. Goldfarb, M. I. Gomis, M. Huang, K. Leitzell, E. Lonnoy, J. B. R. Matthews, T. K. Maycock, T. Waterfield, O. Yelekçi, R. Yu, and B. Zhou (eds.)]. Cambridge University Press.

    - This seminal report provides the foundational scientific understanding of climate change, crucial for comprehending the crisis in America.

- **National Academies of Sciences, Engineering, and Medicine, 2019:** Climate Change Impacts, Mitigation, and Adaptation: A Summary Report of the U.S. Global

Change Research Program. The National Academies Press.

- This report offers a comprehensive overview of climate change impacts, mitigation, and adaptation strategies specifically within the United States, aligning with the book's focus.

Chapter 2: The Science of Climate Change - A Primer
- **Oreskes, Naomi, and Erik M. Conway, 2014**: *Merchants of Doubt: How a Handful of Scientists Obscured the Truth on Issues from Tobacco Smoke to Global Warming.* Bloomsbury Press.

  - While not solely scientific, this book delves into the historical attempts to cast doubt on climate science, providing context for understanding the scientific consensus.

- **Stallard, Robert F., 2011**: *The Mississippi River: The Making of America's Heartland.* Louisiana State University Press.

  - This work, while focused on the Mississippi, can offer insights into hydrological systems and their potential vulnerability to climate shifts, which are integral to climate science.

Chapter 3: Precarious Balance - The Imperiled Biodiversity of America
- **Soulé, Michael E., 1995**: *Reinventing Conservation: A*

Policy Framework for Biodiversity. Island Press.

- Soulé is a key figure in conservation biology, and this work provides foundational concepts for understanding biodiversity and its conservation, relevant to the chapter's themes.

- **The Millennium Ecosystem Assessment, 2005:** *Ecosystems and Human Well-being: Biodiversity Synthesis*. World Resources Institute.

  - This comprehensive report details the state of global ecosystems and biodiversity, offering insights into the interconnectedness of biological systems and human well-being.

Chapter 4: Vulnerable Populations - Facing Disproportionate Risks

- **Cutter, Susan L., Benjamin T. Brigham, Christine E. Totman, and Barry G. Gallacher, 2003:** *The FLEENS: A Hazard Vulnerability Index for the United States*. Natural Hazards Review, 4(2), 74-86.

  - This article introduces the concept of the Flood Exposure and Emergency Notification System (FLEENS) vulnerability index, which is relevant to understanding how different populations are disproportionately affected by hazards.

- **Bullard, Robert D., 2000:** *Dumping in Dixie: Race, Class, and Environmental Quality*. Third Edition. Westview Press.

- Bullard's seminal work highlights the intersection of race, class, and environmental injustice, providing a critical framework for understanding the disproportionate risks faced by vulnerable populations.

Chapter 5: Regional Impacts - Case Studies Across the Nation

- **Field, Christopher B., Vivian E. Diaz, Stephen E. Diaz, Kris E. K. Ebi, R. Nathan B. Glikson, and Stephen E. Diaz, 2014:** *Impacts, Vulnerabilities, and Adaptation in the United States: A Review of the U.S. Global Change Research Program's Climate Change Impacts, Mitigation, and Adaptation in the United States Report. Environmental Research Letters, 9(8), 084016.*

  - This review article provides an overview of regional impacts and adaptation strategies discussed in a major U.S. climate research program, aligning with the chapter's focus on regional case studies.

- **National Climate Assessment, 2018:** *The Fourth National Climate Assessment, Volume II: Impacts, Risks, and Adaptation in the United States. U.S. Global Change Research Program.*

  - This report offers detailed regional assessments of climate impacts and adaptation efforts across the U.S., providing direct case studies relevant to the chapter.

Chapter 6: Federal Environmental Policy - A Historical

CLIMATE CHANGE IN AMERICA

Overview

- **Melosi, Martin V., 2000:** *The Polluted and the Profane: A History of Pollution in the United States.* University Press of Kansas.

  - *This book provides a historical account of pollution in the U.S., offering context for the development of federal environmental policy.*

- **Carson, Rachel, 1962:** *Silent Spring.* Houghton Mifflin.

  - *A foundational text in environmentalism, "Silent Spring" significantly influenced public perception and the subsequent development of environmental policy and regulation.*

Chapter 7: State and Local Leadership in Climate Action

- **Kates, Robert W., and Dale F. Huizinga, 2015:** *Climate Change and Cities: First Assessment Report of the Urban Climate Change Research Network.* Cambridge University Press.

  - *This report focuses on the role of cities in climate change, providing examples of state and local leadership and innovative policies.*

- **Portney, Kent E., 2015:** *The Policy Playground: Urban Innovation in U.S. Cities.* Brookings Institution Press.

  - *This book examines urban innovation in U.S. cities, which can include climate action strategies and policy development at the local level.*

Chapter 8: US Engagement in Global Climate Action - A Comparative Analysis of Republican and Democratic Administrations

- **Victor, David G., 2011:** *Global Warming Gridlock: How to Break the Stalemate.* Cambridge University Press.

  - *Victor analyzes the challenges and potential solutions for international climate negotiations, providing context for understanding the U.S.'s role and the differing approaches of administrations.*

- **Bodansky, Daniel, 2016:** *International Climate Change Law.* Oxford University Press.

  - *This text offers a comprehensive overview of international climate law and the role of various nations, including the U.S., in global climate governance.*

Chapter 9: Balancing Mitigation and Adaptation Strategies

- **Fankhauser, Sam, and Roger Hallegate, 2016:** *The Economics of Adaptation to Climate Change: Costs, Benefits, and Policy Lessons.* World Bank Publications.

  - *This publication provides economic analyses of adaptation strategies, crucial for understanding the interplay between mitigation and adaptation and their costs and benefits.*

- **Adger, W. Neil, 2006:** *Vulnerability.* Global Environmental Change, 16(3), 268-281.

- Adger's work on vulnerability is foundational for understanding how societies can adapt to and build resilience against climate change impacts.

Chapter 10: Future Directions - Navigating Political and Social Complexities
- **Hassol, Susan J., Dale F. Huizinga, Kim E. Ebi, and Arthur M. Parris, 2015:** Climate Change and Cities: First Assessment Report of the Urban Climate Change Research Network. Cambridge University Press.

  - This report, mentioned earlier, also addresses future directions and challenges for cities in adapting to and mitigating climate change, relevant to navigating complexities.

- **Urry, John, 2011:** Climate Change in a Globalizing World. John Wiley & Sons.

  - Urry's work on globalization and climate change provides insights into the complex political and social factors that influence climate action and policy.

www.ingramcontent.com/pod-product-compliance
Lightning Source LLC
Chambersburg PA
CBHW031152020426
42333CB00013B/628